BT COMMUNICATIONS TECHNOLOGY SERIES 6

Telecommunications Network Modelling, Planning and Design

Other volumes in this series:

Telecommunications Network Modelling, Planning and Design

Edited by
Sharon Evans

The Institution of Electrical Engineers

Published by: The Institution of Electrical Engineers, London,
United Kingdom

British Library Cataloguing in Publication Data

A catalogue record for this product is available from the British Library

ISBN 0 86341 323 4

Typeset in the UK by Bowne Global Solutions Ltd, Ipswich
Printed in the UK by T J International, Padstow, Cornwall

D
621·385
TEL

CONTENTS

PREFACE

When people talk about network modelling, the first thing that often springs to mind is a computerised 'map' of the network showing its geographical layout and its traffic flows. And indeed this is one of the many aspects of communications network modelling. But there are many more network modelling disciplines, each addressing the many questions posed by systems and solutions designers.

As it is often the case that one aspect that is being modelled overlaps with another, individual models and analysis cannot be considered in isolation. For example, a network solutions designer has two options — one involves a centralised network, the other utilises a distributed one. From a network performance perspective it might be better to design a centralised network, but from a return on investment viewpoint the decentralised network may offer lower costs. And so models today are designed to be flexible and able to cope with a variety of 'what if' scenarios — a level of sensitivity analysis can then be incorporated and the optimum solution reached.

This very flexibility results in ever larger volumes of data being generated, and, without the aid of continually improving modelling techniques and tools, we would struggle to make sense of that data. The modelling tools help us to analyse different situations, and the outputs are often used as part of a design debate rather than a definitive answer. Increasingly, solution designers work collaboratively with a variety of specialist modellers to meet the ever more sophisticated requirements of customers.

This book offers an insight into some of the modelling disciplines utilised in the design of modern day communications networks.

Sharon Evans
Business Modelling, BT Exact
sharon.m.evans@bt.com

INTRODUCTION

The preface has talked in general terms about modelling concepts and the reasons why models exist. But, as you may know, there are many fields of modelling and this book sets out to introduce you to a selection of communications network modelling disciplines. It has been organised in such a way that each area has its own chapter and, while these can be read individually, the designer should attempt to keep the 'bigger picture' in mind.

The opening chapter describes BT's Utilisator tool and how the outputs have provided solutions not only to network design questions but also to architectural issues.

Chapter 2 moves on to consider a different aspect of network modelling — how to design a network that is robust, resilient and survivable. Networks are now an integral part of a company's infrastructure and recent catastrophic events have demonstrated how much a business comes to rely on the resilience of its networks.

This leads us on to the question of capacity (which is considered in Chapter 3) — how to design and plan a network that has neither too little nor too much (wasted) capacity, a subject which will be familiar to anyone who has been involved with designing a network.

Until now we have looked at how the network should be planned and designed. We have seen modelling techniques that aid in that process. Let us now turn to a network already deployed — the PSTN (public switched telephone network). It has been around for a long time now, and, like most things, can deteriorate with age. In order to ensure that any deterioration does not result in a loss of service, it is better to examine the condition of the network before problems are encountered. Chapter 4 describes a Bayesian network datamining approach to modelling this problem in such a way that deteriorating plant can be identified in good time.

And now on to something rather different. Chapter 5 takes a look at the emergence of unplanned topological traits in an SDH network. Chapter 6 also looks at some different network traits — but this time, in connection with electromagnetic emissions; not something which may immediately spring to mind, but none the less important.

Moving on from modelling of the network itself, Chapter 7 explains how the randomness of both the input and the environment can be mathematically modelled and analysed to improve the system performance of a network.

We now leave behind the network with its various architectures, properties and traits, and move on in Chapter 8 to a fundamental business issue — revenue and cost and how modelling can help to minimise system expenditure.

Chapter 9 moves into the realm of radio resource management for the delivery of multimedia services and describes how quality of service simulation models utilising different algorithms can lead to improved performance.

Now let's look more to the future. Chapter 10 shows how nature can inspire us to solve problems and come up with innovative solutions — not modelling in the traditional sense but a clever way of using nature's real-life models to develop technology, essential in the telecommunications world.

Our last chapter — but no less important for that — looks at security. The solution has been designed, and everything that can be modelled in pursuit of a first rate solution has been modelled. But even the most optimally tuned network needs to be secured against deliberate attack and/or accidental failure. Chapter 11 describes proposals modelled on nature's own immune system.

Finally, I would like to thank all the authors and reviewers for their valuable contributions towards this book and for willingly sharing their knowledge and experiences. I have thoroughly enjoyed learning about those modelling disciplines outside my own area, and I hope you also have pleasure in reading this anthology.

Sharon Evans
Business Modelling, BT Exact
sharon.m.evans@bt.com

CONTRIBUTORS

S Abraham, Mahindra BT, Ipswich

C P Botham, Broadband Network Optimisation, BT Exact, Adastral Park

M Brownlie, Optical Network Design, BT Exact, Adastral Park

D J Carpenter, Business Assurance Solutions, BT Exact, Adastral Park

S Devadhar, Mahindra BT, Ipswich

A M Elvidge, Business Modelling, BT Exact, Adastral Park

P Gaynord, Broadband Network Optimisation, BT Exact, Adastral Park

D J Hand, Professor of Statistics, Imperial College, London

A Hastie, Transport Network Design, BT Exact, Adastral Park

N Hayman, Transport Network Design, BT Exact, Adastral Park

D Johnson, Transport Architecture and Design, BT Exact, Adastral Park

N W Macfadyen, Network Performance Engineering, BT Exact, Adastral Park

J Martucci, Business Modelling, BT Exact, London

C D O'Shea, Broadband Network Optimisation, BT Exact, Adastral Park

A Rai, Mahindra BT, Ipswich

L Sacks, Lecturer in Electrical and Electronic Engineering, University College, London

F Saffre, Future Technology Research, BT Exact, Adastral Park

P Shekhar, Mahindra BT, Ipswich

J Spencer, Department of Electrical and Electronic Engineering, University College, London

R Tateson, Future Technology Research, BT Exact, Adastral Park

A Tsiaparas, formerly Broadband Network Engineering, BT Exact, Adastral Park

D Yearling, formerly Complexity Research Statistics, BT Exact, Adastral Park

1

TRANSPORT NETWORK LIFE-CYCLE MODELLING

M Brownlie

1.1   Introduction

From around 1998 onwards, an increasing number of organisations, operators and joint ventures were building vast pan-European networks. The drivers for such growth were relatively straightforward: European deregulation had opened up hitherto inaccessible markets and prices for high-bandwidth network technologies were becoming cost effective, as demand for high-bandwidth services increased. In such conditions the business case for the rapid deployment of large-scale optical dense wavelength division multiplexing (DWDM) and synchronous digital hierarchy (SDH) networks across Europe was irresistible. At its height, Europe boasted in excess of 25 such networks, at varying degrees of development and scale.

All these new network operators had something in common. They were all effectively building new networks on a 'greenfield' basis, and were developing the teams and tools to build and manage their networks almost from scratch. One such operator was BT's pan-European network deployment, then known as Farland and now called Transborder Pan-European Network (TPEN).

Established on the lines of a start-up, the Farland team's blueprint was based on small interactive units that could work quickly and efficiently in order to build the network they needed, unrestricted by legacy equipment. In order to capture the market most effectively, Farland rolled out the first 10 Gbit/s pan-European network in May 1999. The network started out thinly spread in order to capture the majority of initial demands. It then quickly grew to increase its coverage in new areas and to reinforce coverage in existing areas that would allow it to meet the demanding service level agreements (SLAs) that it had set with its customers.

The Farland network consists of high-capacity, point-to-point, DWDM line systems, interconnecting major population centres across Europe, offering either 16 or 32 × 10 Gbit/s channels per fibre. Overlaid on this infrastructure are a number of SDH rings that have a multiplex section – shared protection ring (MS-SPRing) protection scheme. This 'SPRings over DWDM' approach is commonplace among the pan-European network operators as it combines high capacity, with resilience and operational simplicity.

Like Farland, other networks grew to support more traffic from more European points of presence (EPoPs). These expanding organisations found themselves facing similar issues to those of the more established operator. Many of these issues were associated with the creation and enlargement of teams within the organisation and particularly with the management of the information that was being created, transferred and interpreted between them. Indeed, one possible consequence of a pan-European network is that there are many disparate teams that not only have different functions and responsibilities, but also have many variations in working practices and languages. Similarly, many issues could arise from the sheer scale and complexity of the network topology, its interconnectivity, and its usage. This could manifest itself into a lack of overall insight and clarity regarding the state of the network and consequently any confident drive and direction, that the network originally had, could be lost.

One of the initial methods BT employed in order to prevent these issues from arising was to develop a single repository for network information that presented the relevant network information in different ways to suit the user. This tool was known as the 'Utilisator'.

In the space of around five years, BT's pan-European network (as did many of its competitors) passed through a number of distinct phases. The first was a concerted effort to reach and connect as many customers as possible in order to create initial revenues. This was followed by a more controlled expansion to achieve an optimum balance between network investment and customer revenues. When it became evident that bandwidth demands were falling short of forecasts, the business focus turned to the maximisation of the return on investment in the network by increasing network efficiency and minimising operational spending. Throughout all of these phases, it was vital to have a clear, unambiguous and accurate appreciation of the network — its elements, its connectivity, its utilisation/efficiency and its potential. The Utilisator tool was central to this understanding and has proved invaluable to BT in the functionality that it provides.

What follows in this chapter is a description of the Utilisator tool from the point of view of the people and teams that use the tool the most. It describes the information upon which the tool draws to provide its outputs, the views and direct outputs that result from using the tool, and, perhaps most importantly, how this resultant information can be used within the business to facilitate decision making.

1.2   Creating a Profit-Driven Network

Shorn of all hype and over-optimism, today's network operators need to focus on real profit targets based on realistic revenue opportunities and sound cost management. However, a network operator in a dynamic market-place, has difficulty in defining the metrics by which the network is measured and then identifying the sources of revenue within the network and the areas where money is being unwisely spent.

The desire to maximise the revenue potential of the network while minimising expenditure leads to conflicts and compromises particularly with respect to expansion or upgrade plans for the network.

In order to maintain the correct balance between these conflicting requirements and to create and maintain a profit-driven network an operator must ensure that the four main points below are achieved.

- Minimise operational, systems and support expenditure:

    — align goals and objectives across teams;

    — provide a common information platform;

    — ensure all processes are co-ordinated and streamlined and have the appropriate support systems.

- Maximise network revenue potential:

    — understand the network topology;

    — track component inventory and location;

    — understand the connectivity relationships of network elements;

    — define and frequently monitor network utilisation;

    — optimise network element usage based on customer traffic demands.

- Minimise network operational and capital expenditure:

    — calculate where and when new equipment will be necessary;

    — optimise the architecture and network design to provide services to the largest number of customers at minimum cost;

    — understand the advantages/disadvantages of new network architectures and methodologies.

- Grow revenue from new services:

    — optimise network architectures to minimise delay and maximise reliability;

    — pursue new technologies that enable new and improved services.

The rest of this chapter will develop the ideas listed above and show, where appropriate, how BT has harnessed Utilisator's breadth and depth of functionality to allow them to achieve these goals in order to stay competitive in the European market-place.

1.3   Minimise Operational, Systems and Support Expenditure

Large networks generally need large, well co-ordinated teams in order to monitor and manipulate all the various and interrelated aspects of the network. It is sometimes too easy to lose track of developments, overlook important information or have multiple teams duplicating work effort. Utilisator can be used as a common software application that can keep teams informed of network status thus allowing them to remain focused on their individual objectives. For example, a network may be supported by an array of teams such as sales and marketing, operations, low-level design and high-level strategic planning. Utilisator can be used as the common application that interconnects these teams together by incorporating it into the processes that these teams use to interact with each other. In such an environment Utilisator helps to minimise operational, systems and support expenditure. This idea is expanded upon in the following example.

Figure 1.1 demonstrates how the Utilisator tool can be central to the information flow between various groups within the organisation. In this example, the sales and marketing teams produce the forecast traffic matrix that the planning team uses as an input to Utilisator in order to model the growth of the network. Conversely the sales team could look at the latest network file on Utilisator, that was produced by the planning team, to monitor capacity take-up and use the statistics to provide price-weighted service offerings based on routes and/or locations that are over- or under-utilised. The low-level design team could also use Utilisator as clarification of any build they have recently closed off, and operations could use Utilisator to

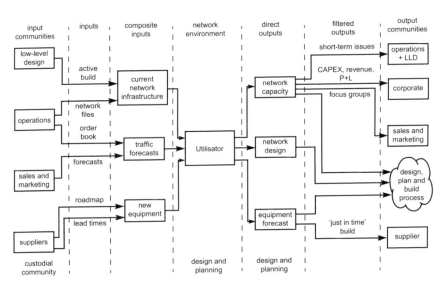

Fig 1.1   Capacity planning process diagram.

retrieve customer statistics, send out planned works notifications to customers and monitor circuits for poor routes, high latencies and/or low availability. For further information on Utilisator's most beneficial features, see the Appendix at the end of this chapter. Incorporating Utilisator into the business processes could help streamline the business in general and provide a unifying source to reference the network across the business. Different streams of this process would be applicable depending upon the format and structure of the organisation and what particular type of modelling scenario was being carried out at any one time.

Consider the information flow shown in Fig 1.1 in more detail. Before Utilisator can perform any modelling work, information has to be gathered from across different areas within the organisation. This is shown in the first column — input communities. Each of these communities can provide input data that falls into one of three distinct composite input categories. These categories are current network infrastructure, traffic forecasts and new equipment. This input data can then be amalgamated and structured in such a way as to be easily incorporated into Utilisator.

Current network infrastructure gathers the relevant network files from the network management system (NMS) in co-operation with the operations department. In addition to this (if required) any current build activities carried out by the low-level design team can be captured as part of this data capture. Traffic forecasts comprise a consolidated forecast list from any remaining ordered forecasts not accepted in the NMS from operations and any customer forecast lists from sales and marketing. The new equipment input to Utilisator would be mostly applicable to strategic network modelling exercises. It would generally be related to additional functionality that would allow Utilisator to accurately model new equipment and/or features on the supplier's roadmaps. Under such circumstances the lead times for these releases would have to be taken account of as they may influence when certain types of forecast traffic could be added to the network.

The next stage of the process is to feed the gathered information into Utilisator and process it. In this example, the majority of control has been given to the design and planning department. They are the custodial community that gather in the required inputs to Utilisator, perform the modelling work, and pass on the relevant information to the other teams involved. Another method could be to give each department its own version of Utilisator that contains the functionality it requires to fulfil its role within the organisational structure.

There are three main 'direct outputs' from any modelling activity — network design, network capacity, and equipment forecast. Network design would show the overall design chosen for any modelled network; network capacity would show the overall utilisation of the design based on the input traffic forecast; and finally equipment forecast would detail any additional equipment that would be required to build the designed network. The exact content and format of any filtered outputs obtained from these three direct outputs would be influenced by the type of modelling work that was being carried out. In the example here the custodial community would verify, check and format the direct outputs from Utilisator to the appropriate form for the relevant output communities. If, for example, the objective was to understand the medium-term implications of expected traffic forecasts, a time-dependent input traffic forecast would result in a time-dependent output equipment forecast. This could be used as feedback to the supplier to check against current factory orders and to initialise any additional equipment into the ordering process to ensure deployment at the time specified in the equipment forecast.

For long-term strategic modelling all direct outputs would have to be considered against other models for comparison before any activation of a design, plan and build process for the chosen network upgrade.

This process illustrates how Utilisator can enable its users to communicate more effectively with each other through a common information platform. Each user community benefits from a shared and open working environment. This helps to increase the productivity of all associated parties with the end result of minimising the resource associated with the operational system and its support.

1.4   Maximise Network Revenue Potential

In order to maximise the revenue potential of a network it is necessary to be able to monitor and track capacity take-up regularly and accurately. This will ensure that the network always has enough resources to support new traffic demands and will highlight any re-engineering that the network may require. To successfully achieve this, Utilisator accurately models the current network capacity fill and can output network statistics in an intuitive and user-friendly environment.

To be as accurate as possible in its network inventory and capacity take-up, Utilisator downloads physical network information from the equipment supplier's proprietary network management system. It is assumed that the NMS is the 'master' inventory system that reflects exactly the current build across the whole network. Utilisator downloads all relevant network elements (NEs) and identifies any relevant equipment installed in that NE. It then downloads the connections (links) between those NEs. Finally it incorporates all circuit information that identifies, for each circuit, the specific equipment and SDH time-slot each circuit occupies along its path. This provides enough information in order to display the network (NEs and links) via a graphical user interface (GUI) for easy interaction with the user, as shown in Fig 1.2. The user can select any NE to view its status, fill and the position of all cards in that NE, as shown in Fig 1.3. The user can then easily identify any card, to view the circuits on that card. The user can also view the size (capacity) of any link, how many circuits occupy that link and which time-slot(s) each circuit occupies, as shown in Fig 1.4. Furthermore, information pertaining to a particular circuit on that link can be retrieved by selecting it from a drop-down menu. The circuit path is then highlighted across the network as shown by the thick black line in Fig 1.2.

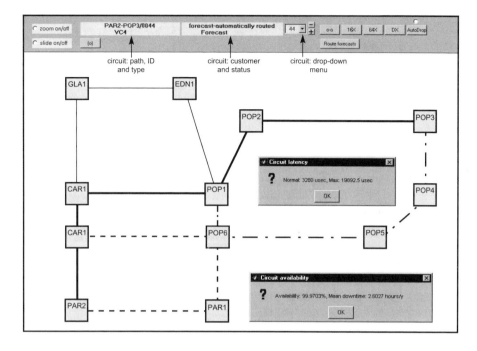

Fig 1.2   Network topology schematic indicating an individual circuit path (thick black line)
with associated latency and availability information.

Utilisator also produces a number of easily digestible network statistics in the
form of reports, graphs and bar charts that can be used to visualise the overall
utilisation of the network. More details of these features are available in the
Appendix.

Such an interface is very intuitive and easy to use. It allows operators to get a real
feel for their network by being able to visualise where all of its components are and,
perhaps more importantly, their associated connectivity. It also allows the same
information to be presented in different ways to suit the user and the purpose of the
query.

Some of the benefits from this functionality include the ability to monitor and
track where capacity 'hot-spots' are forming on the network — allowing the user to
provide card delivery on a 'just-in-time' basis, thus reducing costs from the
elimination of excessive build.

Conversely it could also help maintain high customer circuit-provisioning targets
by ensuring that sufficient interface cards are available at all times to meet demand.
It could also be used to calculate the overall cost of the network and to act as an
early warning system if revenue starts falling unexpectedly against network build
costs.

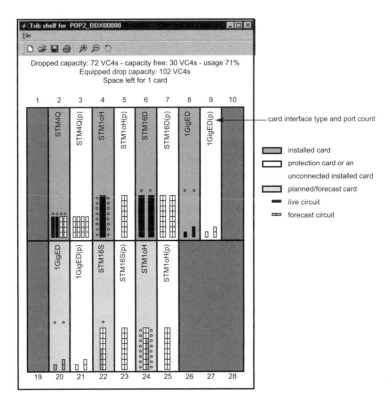

Fig 1.3   Network element view showing types of cards installed, their positions and their utilisation. Forecasted cards/ports can also be indicated.

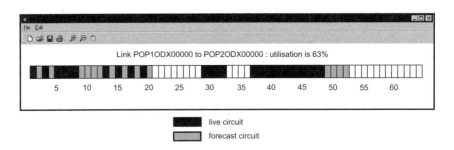

Fig 1.4   Link information showing size and utilisation of that link. This example highlights how VC-4s are distributed within a 10 Gbit/s (STM-64) link.

These features show some of the ways in which Utilisator could be exploited in order to identify revenue-earning opportunities by providing different teams with a simple yet highly advanced, up-to-date and accurate inventory tool.

1.5 Minimise Network Operational and Capital Expenditure

In order for an operator to minimise its network's operational and capital expenditure it must minimise its field engineer base and ensure that the slimmest network design, using the most appropriate technology, is deployed in the most appropriate places. This is an extremely complex problem that has many subtle interactions and co-dependencies. If these issues could be understood and incorporated into a planning tool, it could greatly de-mystify the planning process, increase confidence in the network designs produced, and allow the work to be carried out by less specialised individuals. In order for BT to get the most out of such a planning tool, it was very important that it should accurately reflect and model its network; it has to do more than just act as an inventory system:

- it has to understand the physical layout of individual NEs as well as their respective functionality;

- it has to understand the network architecture and technology in which the NEs are operating;

- it has to know how customer traffic would route across the network;

- it must be able to understand the impact of new or forecast traffic on network design, interaction and efficiency.

Some of the main features that BT wanted to take account of, and which have been incorporated into Utilisator to achieve these goals in order to ultimately reduce the network's operational and capital expenditure, are as follows.

- Constraint-based routing

  The NMS routes circuits along the shortest cost path between two points. Utilisator ensures that the link costs inherent in its own model are the same as those in the NMS. This ensures that all capacity forecasts can be made in the confidence that circuits would be routed by the NMS in the same manner.

- No time-slot interchange (TSI)

  In general, the MS-SPRing protection mechanism may only restore traffic that does not use TSI when spanning multiple sections of a ring. This feature leads to potential blocking of new traffic as spare capacity may be stranded on a ring in cases where each span of a ring could support the required bandwidth, but because the free capacity was offered on different time-slots, the traffic cannot be routed. As Utilisator was designed in such a way that forecasted circuits would not allow TSI, accurate capacity limits could be established.

- Dual node interconnect

  Dual node interconnect is an additional protection and routing feature that reduces the number of single points of failure at ring interconnect sites in order to potentially increase a circuit's reliability/availability. The TPEN planning team was keen to understand the impact that potential circuits using this facility would have on both network utilisation and circuit reliability and as a result it was important that the Utilisator tool could model such schemes.

- Circuit interface types

The above three features allow a circuit to be accurately routed across the BT TPEN. It was also important to accurately model the specific interface requirements at each end for individual circuits. An add drop multiplexer (ADM) drop capacity is dependent upon its switch size and the amount of tributary cards that can be added to that ADM. For example, the number and configuration of circuits that can be dropped on a synchronous transport module (STM-1) interface card may be different from that of an STM-4 interface card — or even more subtly, there may be different types of specific interface cards with different drop capabilities.

These issues must be considered and taken into account as the provision of circuits can be significantly delayed if the required interfaces are not present at a site.

The types of card installed in an ADM will generally govern the amount of drop capacity available at that site. This means that for a specific ADM its maximum drop capability will vary depending on the types of tributary cards installed and this, in turn, will be dependent on the customers' interface requirements. When a circuit is routed across the network there has to be a correct interface card in each NE at either end with enough spare capacity to support that circuit type. Utilisator ensures that for all forecasted circuits these interface and capacity constraints are met, and, if not, it will highlight where and how a shortfall exists or it can add the appropriate card automatically if required.

- Headroom

Utilisator provided the BT TPEN planning team with a headroom feature that could be used to determine the amount of usable spare capacity on paths across the network.

To demonstrate the breadth of modelling possibilities that Utilisator can perform, four main network strategy planning areas will be described — short-term, medium-term, long-term and greenfield. These areas, however, are not performed in isolation of the network nor remain part of a theoretical model. To understand the benefits of any modelling work it is important to be able to analyse the results and feed any tangible benefits back into the network as highlighted earlier in section 1.3.

1.5.1   Short-Term Planning

For short-term forecasting the following process is adopted. Within a few minutes a good representation of the capacity constraints and abilities can be ascertained:

- download 'live' network data from NMS;

- add additional 'in progress'/short-term equipment build if desired — this could be any new hardware additions that will be installed in the network during the length of the forecast routing period;

- route customer circuits in order-book/short-term forecasts — this can be achieved in two ways:

  — the first facility is designed to quickly route a handful of circuits only, with the user identifying the end-points of a forecasted circuit and the tool selecting the best route between them (this route can be overridden manually by the user if desired);

  — if there are a large number of circuits forecast, the user can use the second option which is to create a traffic matrix (in a simple text file) specifying various circuit details that can be routed in bulk across the network;

- highlight any additional card build to satisfy short-term forecast, as in many cases the forecast traffic would exceed the capabilities of the current network, hence necessitating new network build — Utilisator can be instructed to either add the new equipment required to support the demand or simply note that a particular demand cannot be routed.

At the end of this process, the planning team is able to decide on the most cost-efficient network build programme based on its experience of forecast demands and from priorities and objectives. It will be able to report to the investment/financial departments either the cost associated with meeting expected demands or the potential revenue lost should such investment not be forthcoming.

1.5.2   Medium-Term Planning

Short-term planning addresses the immediate and pressing customer orders and highlights areas where new cards would be required in existing network elements. For medium-term planning, the same initial process is followed, but the focus centres on whether there is cause to build new equipment capabilities at sites (for example new ADMs or interconnection points) as such activity takes longer to plan and deploy.

The process for medium-term planning is as follows:

- route mid-term forecasts/multiple traffic distributions;

- automatically add additional build to meet requirements, e.g. tributary cards;

- at major build points, interrupt routing process to add appropriate network infrastructure (e.g. ADMs and ring interconnections);

- save various strategies as separate network models — this is so that different scenarios can be examined at a later date to determine the best manner to service the expected medium-term demands.

At this stage the planning team should be able to identify where and when the existing network infrastructure could be nearing exhaustion. Network build programmes could then be initiated.

1.5.3 Long-Term Planning

Long-term planning involves taking both known and potential traffic forecasts and combining them with longer-term trends and internal strategies to indicate how the overall network could develop, expand and evolve over a period of 9-12 months. Such planning is important as significant network build, such as fibre deployments (link augmentations) or the installing of new sites, can take many months to realise.

The process for long-term planning, again, follows similar steps as previously:

- at major build points, interrupt routing process to add appropriate network infrastructure, e.g. new rings, stacked rings, spurs, meshes;

- simulate new products on manufacturer's road map to assess impact on network:

    — replacing current equipment;

    — redesigning current network;

    — enhanced stack design based on actual traffic analysis and/or improved equipment functionality;

    — network expansion.

The long-term plans would feed into network strategy teams in order to provide a coherent deployment plan and to facilitate appropriate business case approval.

1.5.4 'Greenfield' Networks

All of the above scenarios are based on downloading the current network from the NMS as a starting point. Some operators may not be in a position to interact with their NMS directly. For this reason network models can also be built up within Utilisator independently from the NMS. This feature could be used to model an existing network or to model a prospective hypothetical network design. If Utilisator was incorporated into an operator's plan-and-build process, any network upgrades could be reflected 'off-line' within Utilisator.

As an example of this type of activity, consider a network planning team wishing to determine the most suitable network design for a given traffic demand as shown in Fig 1.5.

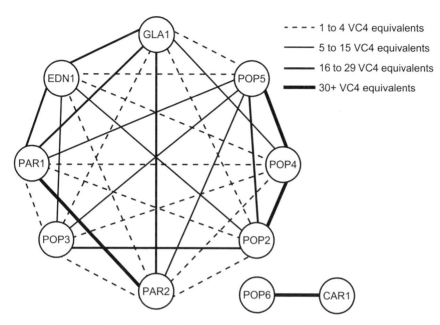

Fig 1.5   Traffic demand for a (hypothetical) proposed network. The thickness of a line is indicative of traffic demand between end-points.

The lines that are indicated in Fig 1.5 denote point-to-point traffic paths; their thickness indicates the amount of circuits between each point-of-presence (PoP) pair. This traffic demand consists of 111 circuits equating to 320 VC4 equivalents made up of a combination of VC4, VC4-2c, VC4-4c, VC4-8c and VC4-16c circuits that are routed across the 10-node network. For the purposes of this example, the planning team is considering two initial design options. The first design is that of a single ring incorporating all PoPs (see Fig 1.6) while the second design is based on a three-ring network (see Fig 1.7). Both network designs were created within Utilisator and those circuit demands that could be supported were routed by the tool whereas those that could not be supported were simply noted (no equipment build was allowed).

An overall impression of the loading of these two network designs can be seen in Figs 1.6 and 1.7, which represent the utilisation of the single ring network and the multi-ring network respectively. By comparing these two figures some observations can be drawn about the two networks. The single ring network (Fig 1.6) has fully utilised one of its links (solid black line) and four more are close to being full. When a link on a ring is used up it is generally necessary to add another ring, if using a SPRing architecture. This could mean a new ten-node ring or an express ring. The advantage of an express ring is that it would be cheaper to deploy as it would only drop traffic at a sub-set of the sites that the 10-node ring dropped at, but this then reduces the flexibility of the ring.

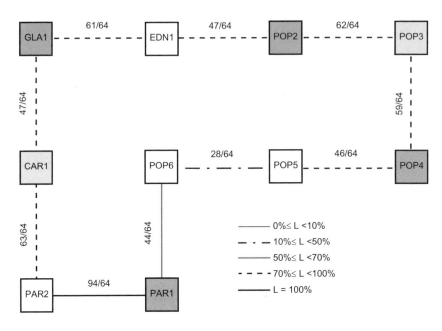

Fig 1.6   Utilisation of single ring network to meet the traffic matrix indicated in Fig 1.5.

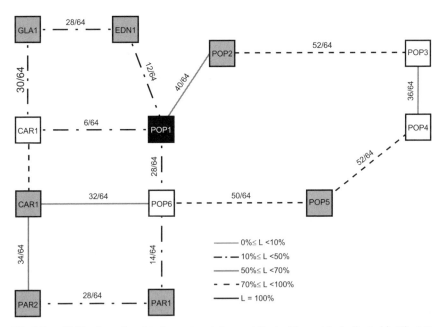

Fig 1.7   Utilisation of multi-ring network to meet the traffic matrix indicated in Fig 1.5.

Both of these options could be modelled in the tool in order to understand what impact the design of this second layer would have on the ability to successfully route the rest of the forecast traffic.

Examining the multi-ring solution (Fig 1.7), it can be noted that one of the constituent rings is close to exhaustion. This network will shortly have to add another 6-node ring that could connect into the other two rings, which still have a lot of spare capacity. It is not really large enough to consider the option of an express ring.

The specific result of this process shows that the single ring network routed 89 circuits that equated to 211 VC4 equivalents and the multi-ring network routed 96 circuits which equates to 236 VC4 equivalents.

From the above results, the multi-ring network has routed more circuits and has more spare capacity than the single ring design. There are two reasons for this outcome.

The first is that there are more routing options, and hence bandwidth, available in the multi-ring network. The second is that the multi-ring network was able to route more of the concatenated circuit traffic because having multiple rings meant that there was less of a possibility of having stranded capacity as a result of the lack of time-slot interchange.

To summarise, the single ring network would be cheaper to install as it needs less equipment but it is perhaps less desirable in terms of routing options, upgrade paths and overall flexibility. The multi-ring network would cost more initially as it requires more equipment and fibre infrastructure but it can accommodate more circuits, has more routing options, is generally more flexible and can be grown incrementally. Further work would have to be carried out over a longer timeframe in order to calculate what impact the design of the layer-2 options would have on the final outcome.

Although such general conclusions can be reached without any detailed modelling, this example shows that Utilisator can provide specific and quantitative answers to specific input information, hence contributing to the decision-making process within the organisation. It provides planners with the evidence required in order to submit a strong business case that will hopefully result in a robust and future-proof network design.

In order to model the take-up of additional traffic on a network it is necessary for any planning tool to behave as closely as possible to the real network that it supports. BT uses Utilisator for its ability to do just this. It allows them to model the TPEN with confidence and at minimum cost by giving them an appreciation of how each circuit will effect the network capacity, ultimately allowing them to know where and when any new equipment build will be necessary. Utilisator allows this to be done far more quickly, more accurately and with fewer resources than could be achieved manually.

This provides additional cost savings in terms of reduced time, resources and network planning errors. BT exploits these benefits in order to help them minimise the operational and capital expenditures of the TPEN.

1.6   Grow Revenue from New Services

The planning and visualisation features above all help to improve the services that the TPEN provides to its customers. Being able to optimise the network, and anticipate where customer demand will occur, ensures a short turn-round in providing services that help keep the order book as short as possible.

Being able to design hypothetical models, based on future trends in network design, helps to investigate new possible service offerings or the reduction in price of current offerings.

1.6.1   Availability and Latency

One incentive that can attract customers is the level of guaranteed availability that is quoted for that service and latency guarantees for time-critical services. As well as designing for optimum utilisation, Utilisator can also carry out availability and latency calculations on any circuit in the network as shown earlier in Fig 1.2. This feature can be used to find the correct balance between high service guarantees and cost-effective network designs.

1.6.2   Planned Works Notification

Customers invariably demand the highest levels of availability for their circuits. Under some circumstances this could be put at risk through essential planned works on the network. Being able to quickly alert people about the details of these planned works and what circuits may be affected is very useful information to the customer. Utilisator can be used to select any NEs, links or PoPs that will be affected by the planned works, and to provide a specific report containing any relevant information for each customer that may be affected.

1.6.3   New Technology Support

As network technology evolves, so do the services that can be offered to the customers. Utilisator can be adapted to accurately reflect the functionality and design implications of new technology deployed in the network and can, in association with the functionality described above, provide firm evidence to support (or otherwise) the provision of such services.

1.7   Future Developments

Utilisator is continually updated to maintain its accurate representation of the network. Work is carried out in conjunction with both the supplier and operator to track any new equipment and strategies planned for the network. This ensures timely and relevant releases of any upgrades that are required.

Additional content and usability for the user interface (GUI) are also improved when necessary through close interaction and understanding of BT's requirements and preferences.

There are also more long-term, strategic developments that may be incorporated incrementally into Utilisator in order for it to continue to be relevant to BT and others. Some of these features are outlined below.

- Integration with network management system

  A major goal of Utilisator is to interact through open interfaces with the NMS. An advantage of this can be seen by the following example. Utilisator's quick and easy forecasting functionality could be taken advantage of more directly by the operations team in the network management centre. Forecast circuits can be routed on a least-cost basis on Utilisator and the appropriate path and bandwidth could also be reserved on the network.

  When the circuit is provisioned on the network following the path that was set up on Utilisator its status would change from 'forecast' to 'provisioned' and this information would also be feedback to Utilisator. This should improve the operations team's ability to manage customer demands and expectations by quickly assessing an order's status and lead-time.

- Convergence of other network layers

  Currently, Utilisator's traffic-handling capabilities relate to SDH, VC4 and VC4-nc (where n is equal to 2, 4, 8, 16, 64) demands and the relevant equipment that supports those demands. In the future, other traffic demands will be considered such as wavelengths and sub-VC4 demands along with the relevant NEs that support those traffic types. Indeed, non-SDH-based services may be considered for inclusion should service demand be realised.

- Convergence of other supplier's equipment and NMSs

  One way in which operators can maintain a competitive edge in their market-place is to have multiple suppliers providing 'best-in-class' equipment. This ensures that the suppliers are innovative in their network offerings and allows the operator to have some financial leverage in any dealings that may take place.

  In the future, Utilisator could be able to reflect this business model by incorporating other network manufacturer's equipment into its workspace. It may also be able to interface directly with these suppliers' management systems and facilitate certain communications between them. This would allow the management of services that cross management and supplier domains in a seamless manner as perceived by the operations and planning teams.

Fulfilling these proposed development points along with other considerations will add to Utilisator's existing functionality and usefulness and make it more effective in supporting an operator's ability to harness all its resources within the organisation.

1.8   Summary

The Utilisator network inventory and capacity-management planning tool was developed by BT to support its pan-European network.

It captures the 'live' network inventory information directly from the supplier's NMS in order to ensure accuracy of data and graphically displays this information and issues meaningful reports in a user-friendly environment. Most importantly, it is a capacity-planning product that accurately reflects the hardware and software features of the network it represents.

Utilisator allows the TPEN planning team to predict how their network would be affected by additional traffic demands by being as accurate as possible in the way it routes and delivers those demands.

Utilisator has become an integral part of the way the TPEN operates. It plays a pivotal role in major strategic and future development processes by allowing the TPEN team to manage, monitor and control their network costs and revenues. Its ability to present the information it contains in a useful and intuitive manner also allows it to be accepted by a large user community who can be unified and co-ordinated under its umbrella.

For these reasons Utilisator has an important part to play in helping the TPEN to operate successfully in any market environment by minimising its operational and capital expenditure and maximising its revenue earning potential.

Put simply, Utilisator is more valuable than the sum of its individual parts. It consistently meets the expectations of the many different users who rely on it to provide them with a clear and accurate representation of the network's current status and future possibilities.

Appendix

1A *The Main Features of Utilisator*

Function type	Function	Description
Utilisation	Plot network capacity	Show colour coded network utilisation
	Inter-ring capacity	Bar charts of augmented interconnect capacity
	Ring capacity	Bar charts of augmented ring capacity
	Ring time-slot map	Bar charts of actual ring capacity
	Popular paths bar	Chart of most popular routes
	Inter/intra-ring traffic	Bar charts of general traffic statistics of rings
Build	Add equipment	Add ADM
	Add drop	Add terminating tributary drop cards
	Add link	Add ring aggregate or tributary ring interconnect link
	Add PoP	Add new site
	Add ADM into ring	Cut-in ADM into an existing ring
	Delete link	Delete Link
	Delete equipment	Delete ADM
Circuits	Route forecasts	Automatically route forecasts by importing traffic matrix from text file
	Search	By: type, customer, ID,
	Calculate delays	Calculate the latency of a circuit
	Calculate availability	Calculate the availability of a circuit
	Print circuit path	Display circuit connectivity across whole path
Planned works	Planned works notification	Select the node(s), link(s), PoP(s) affected by planned works and output a list of circuits and the associated customers that will be affected
View	View network	GUI display
	View link	Display link capacity and utilisation
	View ADM	Display shelf view of ADM showing all trib cards and card utilisation
	Select layer(s)	Select what equipment and/or rings to view
	View circuit path	View circuit path across network and circuit information for circuit selected Interactive
Interactive GUI	Move 'elements'	User can move all PoPs, nodes and links
	Zoom	User can zoom in to see more detail

2

ADVANCED MODELLING TECHNIQUES FOR DESIGNING SURVIVABLE TELECOMMUNICATIONS NETWORKS

C P Botham, N Hayman, A Tsiaparas and P Gaynord

2.1   Introduction

As a key enabler of 'broadband Britain', near-future multimedia communications will require high-capacity networks realised through optical wavelength division multiplexing (WDM) technology. Such systems have the potential to cater for enormous numbers of customers simultaneously, making fast and efficient restoration of service after failure an essential network attribute. Recent world events have also prompted many network and service providers to review their plans and strategies relating to resilience, restoration and disaster recovery on a countrywide and even international scale [1].

Design of resilient networks is a hugely complex process since inefficient designs can result in a combination of unnecessarily high investment, inability to meet customer demands and inadequate service performance. As network size increases, a manual process rapidly becomes unfeasible and automated tools to assist the network planner become essential. This chapter discusses state-of-the-art software tools and algorithms developed by BT Exact for automated topological network design, planning of restoration/resilience capacity, and calculation of end-to-end service availability.

The design challenges [2] associated with automatic network planning are mathematically 'hard' and generally beyond formal optimisation techniques (e.g. linear programming) for realistically sized problems. The tool used by BT relies on iterative heuristics, accurately reflecting the complex structure and guaranteeing wide applicability to a large class of problems. Computational experience has shown that although this procedure is fast and simple, it nevertheless yields solutions of a quality competitive with other much slower procedures. Many extensions of the assumptions are possible without unduly increasing the complexity of the algorithms, and, as the methods themselves are largely technology-independent, they may be applied to a wide variety of network scenarios.

A separate tool models a range of protection and restoration mechanisms in a circuit-based network in response to various failure scenarios. It can audit the resilience of existing networks and optimise the amount of spare capacity for new designs. Again, it is not restricted to any particular technology and can be applied equally well to PDH, SDH, ATM, IP, WDM and even control plane networks.

A third application is a circuit-reliability modelling tool based on Markov techniques. This is capable of representing unprotected and protected paths through network elements and infrastructure using fault data to calculate end-to-end service failure rates and availability. It caters for non-ideal conditions by including factors such as dependent or common-cause failures, fault coverage, the unavailability of protection paths and repair-induced breaks.

A generic network model, representative of the topology and traffic distribution associated with an inter-city transmission network for a large European country, is used to allow the automatic design of mesh and ring networks. Restoration capacity is then planned and optimised for the designs, assuming different resilience strategies. Finally, end-to-end circuit availability calculations are discussed, to illustrate the particular complexities associated with shared restoration schemes.

2.2   Network Model

One cost-effective structure for a resilient network is a mesh-based multi-level hierarchy consisting of a 'core' backbone network and a family of local 'access' networks. The essential inputs to the design process are:

- a matrix of customer traffic requirements;
- candidate sites for nodes;
- equipment (link and node) costs;
- available duct network;
- reliability requirements.

Designated core nodes serve to merge traffic flows so that bandwidth can be used more efficiently, taking advantage of any economies of scale. For modelling, bi-directional traffic and a homogeneous network with identical hardware and software at each node are often assumed, though this is not fundamental. To guarantee a reliable design, the tools may optionally ensure there are two independent (physically diverse) paths between each node pair.

An alternative technique builds a resilient network based on WDM rings, all within the optical layer for fast, easy and immediate recovery. Every working link must then be covered by at least one ring. Upon failure of a link, affected working lightpaths are simply routed in the opposite direction around the ring.

To demonstrate these two architectures, a study was undertaken using a realistic example network. This generic model, developed by BT, is not intended to represent a particular forecast on a particular date, although by scaling the traffic volumes up and/or down, it is possible to represent growth in demand over time. The model is representative of an inter-city transmission network for a large European country and is constructed from:

- actual major transport node locations;

- actual physical layer connectivity, including fibre junction points;

- actual distribution of fibre lengths between nodes;

- actual (non-uniform) traffic patterns.

These factors are particularly important when comparing shared mesh and ring networks. Shared restoration mesh networks minimise the link cost by achieving direct routings for working paths and the highest possible degree of sharing for protection paths. This effect is most significant when links are long (because the savings are proportionately greater), and when the connectivity of the network nodes is high (because a greater degree of sharing of restoration capacity is possible). The traffic pattern is particularly important for ring networks where it is advantageous to be able to fill rings evenly [3].

The network topology is represented by Fig 2.1. There are 119 links, all of which are assumed to be physically separate, with 58 traffic-generating nodes and a further 21 nodes which are required to define the fibre topology. Some nodes are shown with up to 6 diverse routes, whereas in reality there may be short sections close to the nodes where the diversity is reduced by, for example, a common duct running into a building.

The traffic mix, in terms of total bandwidth, is shown in Fig 2.2, where that total is equivalent to over 11 000 STM-1 (155 Mbit/s) demands.

2.3   Design

2.3.1   Mesh

The BT mesh design algorithm generates the topology to best serve customer demand, establishing fibre connectivity between individual nodes subject to the constraints imposed by the available duct network. A very large number of different candidate topologies are explored, searching for an acceptable near-optimum

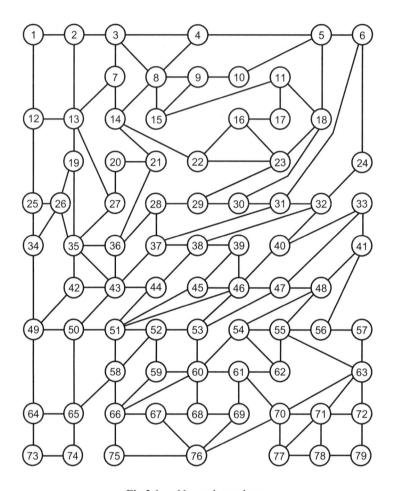

Fig 2.1   Network topology.

solution. The art lies in engineering the search algorithm to operate in a reasonable amount of computer time (typically, minutes rather than hours or days).

While it is conceptually simplest to start from a 'greenfield' site, where none of the network links are known initially, this algorithm is more general. Any links already installed may be labelled as such, with the algorithm subsequently forbidden to delete them. That approach was followed here.

For the particular traffic scenario under consideration, the mesh design algorithm succeeded in reducing the 119 potential links in Fig 2.1 by some 15%, based on a requirement to provide dedicated node and link-diverse back-up routes for each traffic demand, e.g. 1+1 dedicated protection. This represents one of the simplest possible resilience mechanisms available but normally requires greater installed capacity than the more sophisticated approaches discussed later. The corresponding

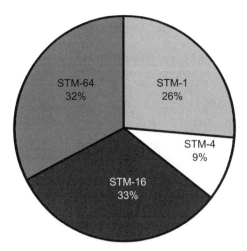

Fig 2.2   Traffic mix in the generic network model.

relative loading on network links and switches is summarised in Figs 2.3 and 2.4 respectively. In general, network capacity is utilised in an efficient manner, with strong correlation between link and switch behaviours, as would be expected.

2.3.2   Ring

The design of survivable all-optical networks based on self-healing WDM rings requires the solution of three sub-problems:

- routing of working lightpaths between node pairs to support traffic demands;
- ring cover of the underlying mesh topology;
- selection of which ring protects which working lightpath.

For the purposes of the present discussion, it should be noted that, as availability of wavelength converters and tuneable transmitters/receivers has been assumed, there are no explicit wavelength-allocation [4] considerations and the issue is purely one of allocating sufficient bandwidth.

The planning approach [5] starts from candidate locations of optical crossconnects, interconnected by the existing duct network, together with demand between each pair of nodes. Every working lightpath is to be protected against single link failure, with typical constraints including:

- maximum ring size (node hops or physical distance) is limited by need for satisfactory restoration time and signal quality;
- maximum number of rings covering a link is limited by network management complexity;
- maximum number of rings crossing a node is limited to control node complexity.

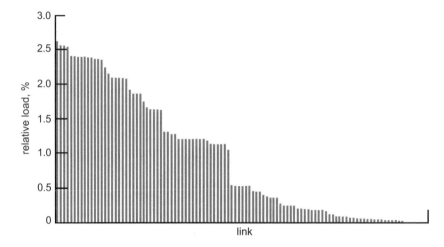

Fig 2.3   Loading on links in mesh network design.

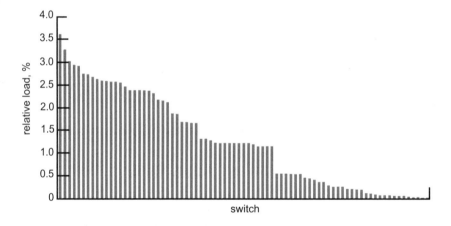

Fig 2.4   Loading on switches in mesh network design.

There are various trade-offs to be considered:

- deploying more rings makes it easier to satisfy the competing constraints but implies more network infrastructure (hence greater installation cost);

- each ring generally traverses a combination of traffic and non-traffic generating locations within the duct network — shorter rings are preferable but should include at least three nodes to provide a meaningful infrastructure for traffic;

- preselection of core nodes affects how large the rings must be to interconnect them, as dictated by the available duct network.

The BT ring design algorithm considers a weighted sum of terms representing each of these conflicting requirements, together with the ring size and coverage limits discussed above. Varying the weights systematically allows a user to choose a 'best' solution according to the desired compromise, with no single network design satisfying all criteria simultaneously.

In the current application, twenty-four rings were identified as 'best' serving the given traffic demands, selected from an initial pool of several hundred candidate rings. With the given pattern of demands, overall resilience can only be provided at the expense of introducing some relatively long rings, but the algorithm is flexible enough to smoothly accommodate this. The profile of traffic load across each ring is shown in Fig 2.5, which is obviously much less uniform than the mesh cases (Figs 2.3 and 2.4), and emphasises the dominance of a relatively small number of rings in this scenario.

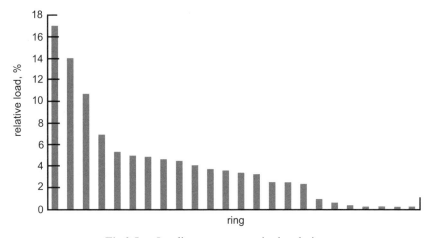

Fig 2.5   Loading on structures in ring design.

2.4   Resilience

A tool called SARC (Simulated Annealing for Restoration Capacity) has been developed by BT to allow the comparison of a range of protection and restoration mechanisms under various failure conditions in a network with an arbitrary topology. It can audit the resilience of existing networks, help in selecting the best resilience mechanism, and optimise spare network capacity.

Networks are constructed from 'nodes', 'subspans' and 'paths', where a node is a flexibility point capable of re-routing blocks of capacity, a subspan is a transmission system connecting two such nodes, and a path is the route a demand takes through the network. This means that, as SARC is not restricted to any particular technology, it can be applied equally well to PDH, SDH, ATM, IP, WDM and even control plane networks. This universality, along with an ability to handle very large models, has allowed BT to perform a variety of studies, including a recurring audit of BT's PDH network (containing several thousand nodes and tens of thousands of links) and cost comparisons of various multilayer disaster recovery strategies for the UK.

The restoration methods that can be modelled in SARC (see Fig 2.6) are:

- adjacent span — the traffic is restored at the system level as closely as possible to the failure via adjacent nodes and spans;

- dynamic path — the traffic is restored at the path level as closely as possible to the failure via adjacent nodes and spans (a different back-up route may be used depending on which part of the original path has failed);

- preplanned path — a pre-set back-up route is assigned for use in restoring/ protecting any failure along the original path (this back-up route will be node and subspan disjoint from the main path).

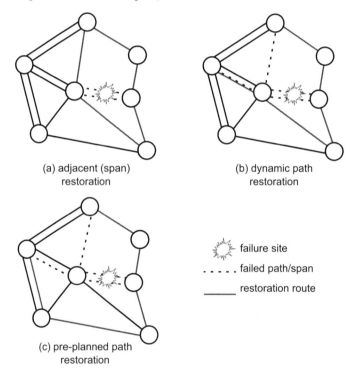

(a) adjacent (span) restoration

(b) dynamic path restoration

(c) pre-planned path restoration

failure site

failed path/span

restoration route

Fig 2.6   SARC restoration options.

As for the failure scenarios, these can be either single subspan (to represent a lone system failure), multiple subspans (to model an entire cable/duct failure) or single and multiple node.

When restoration schemes are being modelled, protection capacity does not have to be dedicated to the restoration of any one span/path, but can be shared between many. If, when using pre-planned path restoration, sharing is not allowed, then the resulting network design has 1+1 dedicated protection. Traffic may be split over more than one restoration route; those back-up paths can either be predefined (for auditing purposes) or left for SARC to choose.

2.4.1   Simulated Annealing

SARC uses a technique called simulated annealing to optimise the cost of providing a specified degree of 'restorability', which is defined as the proportion of working traffic that can be restored following a specified set of network failures. Simulated annealing is derived from an analogy with cooling a fluid to produce a uniform solid crystal structure, which is a state with minimum energy. At high temperatures, atoms in the fluid have enough energy to move around freely. If the fluid is cooled, or annealed, slowly the atoms settle into a perfectly regular crystal structure which has minimum energy. If the metal is cooled too quickly, imperfections are frozen into the structure, which will not then have minimum energy. In simulated annealing, the internal energy of the fluid corresponds to the cost function to be optimised, the positions of atoms in the fluid correspond to the values of variables in the optimisation problem, and the minimum energy state in the fluid equates to an optimal solution of the problem. With difficult optimisation problems, near-optimum rather than global minimum solutions may be found.

SARC can use any solution as a starting point and then small changes to it are proposed; the nature of the small changes depends upon the choice of resilience mechanism. Changes that move the solution closer to the optimal (have lower energy) are always accepted, and, early in the annealing process, most of the solutions that move it further from the optimal are accepted too. This corresponds to a high temperature in the fluid where atoms are free to move away from optimal positions. As time progresses, fewer and fewer of the changes which reduce the level of optimality are accepted, and, if this process is gradual enough, the optimal (minimum energy) solution is reached. Given sufficient time, simulated annealing has the potential to find global optimal solutions or, alternatively, a shorter run time can be traded against less optimality.

2.4.2   1+1 Protection

While it is possible to model a ring-based network in SARC, for simplicity the mesh network design described in section 2.3.1 was used to demonstrate the tool's abilities. Initially, an audit was performed, confirming that 100% restorability in the event of the independent failure of any subspan was possible; this should clearly be the case since the mesh design utilises 1+1 dedicated protection. The spare network capacity required was over 160% of the total working capacity, which is also to be expected since the protection paths have to be node and link diverse from the working paths — hence they will be longer and thus use relatively more network capacity.

2.4.3 Shared Restoration

If the pre-planned protection paths can be shared between different main paths, there are savings to be made with respect to the amount of spare capacity required. This is the fundamental principle behind shared restoration. Judging where and how much (or indeed how little) spare capacity you need is a complex task, usually too complicated for a purely manual approach, which is precisely where SARC comes in.

If the above 1+1 protected mesh design is assumed to have restoration capabilities, e.g. sharing of recovery paths is allowed and re-grooming of traffic can be performed in every node, then SARC can optimise based on the pre-planned stand-by routes already suggested. Letting SARC choose and optimise its own restoration routes (from an extensive list of potential paths), allows the amount of spare capacity required to be further reduced (Fig 2.7).

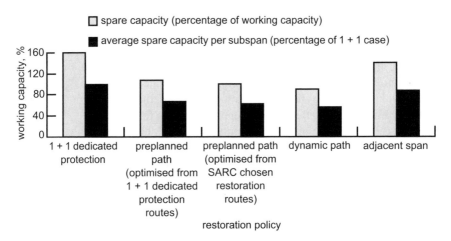

Fig 2.7 Comparison of protection/restoration options for mesh network design.

Both the above options assume preplanned path restoration where recovery paths are end-to-end node and link disjoint from their associated working routes. As mentioned previously, SARC is capable of modelling other restoration schemes, namely dynamic path, where the choice of back-up route depends on which part of the original path has failed, and adjacent span, where working traffic is restored as close as possible to the failure (via adjacent nodes and spans). The results of modelling the network under these restoration conditions, along with the pre-planned path options, are summarised in Fig 2.7. The graph shows the spare capacity required for each approach, as a percentage of the total working network capacity required, and the average loading of spare capacity per subspan, as a percentage of the loading in the 1+1 dedicated protection case.

This high level view of the 'best' restoration strategy does not tell the whole story, but certain conclusions can be drawn.

Although the preplanned path option allows (relatively) simple management and control of restoration since the back-up routes are known before any failure occurs, resulting in 'fast' restoration in the order of 100 ms being possible, it may not generate the cheapest transmission network design due to the level of spare capacity required. Also, it only functions truly well if the record of working and restoration routes is accurate, up to date and valid, so that an unavailable or non-existent recovery path is never used. In all restoration schemes, managing appropriately deployed spare capacity can be a time-consuming and computationally intensive process. Decisions must be made on the frequency of audits and the level of overhead maintained that is not only sufficient to handle failure scenarios but also commercially justifiable, especially if there are demands for that spare capacity to be utilised for working traffic.

A direct consequence of letting SARC choose and optimise its own restoration routes, other than greater sharing of spare capacity (and hence a reduction in the total amount needed and a more even spread of it), is an increase in the length of the average restoration path. This is illustrated in Fig 2.8 by a demand between nodes 35 and 32 from the generic network model (Fig 2.1).

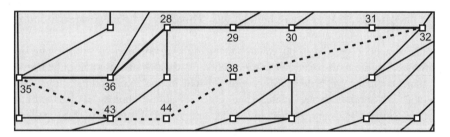

(a)   Circuit with 1 + 1 dedicated protection.

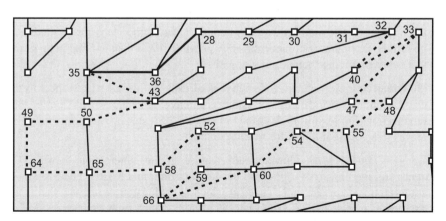

(b)   Circuit with optimised preplanned shared restoration.

Fig 2.8   Back-up paths.

The lengthening of restoration routes can have a serious impact when considering purely optical networks because signal degradation comes increasingly into play. As a consequence, some back-up paths could now require intermediate electrical (3R) regeneration, which can be expensive when required on large numbers of paths.

The dynamic path scheme offers an advantage over a preplanned path since it tends to have shorter restoration routes and hence can function with a lower spare capacity overhead, due in part to the better spread of the required spare bandwidth. This is ultimately determined by the number and diversity of the underlying transmission systems; so (as in this case), if there is not a fully meshed network, the difference compared with end-to-end diverse back-up paths can be small. Dynamic path has to be able to restore quickly after failures, e.g. by deciding what back-up path(s) should be used for the specific incident, to match the performance of the preplanned path method. There is an ongoing discussion about just how fast protection and restoration mechanisms need to be when recovering traffic before the client network actually detects a failure [6]. With protocols such as ATM and IP, provided the break is sufficiently short that the data layer does not start re-configuring virtual paths and/or updating routing tables, outages many times longer than the oft-quoted 50 ms may be tolerable. This does, of course, depend entirely upon the client applications.

The adjacent span method relies on the bulk restoration of entire subspans. Compared to the path-based restoration schemes, this results in much higher levels of spare capacity and a level of system fill that is less than optimal, since large volumes of bandwidth are switched together as single chunks. It does produce slightly shorter back-up routes and saves on switch costs associated with re-grooming a multitude of individual paths.

The amount of spare capacity required is high in some of the shared restoration cases (see Fig 2.7). There are two main reasons for this:

- the working traffic routings were kept constant across the scenarios — there may be further sharing of restoration capacity possible if less than optimal main routes were chosen, giving 'better' pairs of paths;

- the underlying mesh network design was already optimised with respect to the cost of routing the traffic by not using certain available duct routes (that is what the mesh design algorithm described in section 2.3.1 does) — consequently this leaves less potential routes for restoration paths.

Compensating for the above would require greater interaction between the designer and the tools (both the mesh design algorithm and SARC) and a series of (many) iterations, but due to the speed at which the software can operate, that is not as onerous as it may appear. The final decision on which protection or restoration policy to adopt is usually cost driven, more so under current economic conditions than ever. As SARC allows fast and accurate investigation of many options, a network designer should quickly be able to make informed recommendations on which scheme is 'best'.

2.5   End-to-End Service Availability

Based on Markov reliability modelling techniques, a circuit-availability tool has been developed by BT that can calculate the availability of unprotected and protected paths through network elements and infrastructure using appropriate fault data and repair times. It assumes certain non-perfect conditions by having factors such as dependent or common-cause failures, fault coverage, protection path unavailability and repair-induced breaks built into the tool. These aspects are explained later, after a brief description of the Markov approach to reliability modelling.

2.5.1   Markov Reliability Modelling

The Markov technique is a widely recognised method for reliability modelling. It uses the concept of state analysis to model the behaviour of a system as it progressively fails from an initial working situation. Probabilities are used to define the transitions between the possible states of a system; they are determined from the failure rates and repair rates associated with the field replaceable units (FRUs) of which the system is comprised. The transition probabilities act as coefficients in a set of differential equations which, when solved using a suitable method (such as Laplace transforms), give the probability of the system being in any particular state at a given time. Once these state probabilities have been determined, it is then possible to calculate other system parameters, such as failure rate, availability, etc. More detailed explanations and derivations can be found elsewhere [7, 8].

For Markov modelling to be valid, there are normally two main criteria to be considered:

- all transition times must be exponential;

- transition probabilities depend only on the present state of the system.

Since equipment deployment in an evolving network is generally spread over a number of years, giving a reasonable distribution to the age of in-service kit, and purchases by major operators tend to be in large quantities, variations in failure rates over time get smoothed out and any statistical variation of in-service reliability is greatly reduced. Also, service providers are primarily concerned with average behaviour over the lifetime of the equipment; this can be anything up to 15 years, which is much greater than the period of any 'infant mortality'. This implies that the probability of the equipment being in any state will be approximately constant with time, allowing steady state solutions to the differential equations to be considered.

Although there are some situations where the transition probabilities do not only depend on the present state of system, e.g. a failure induced by external events, it can be argued that such incidents can be treated separately from the main analysis. Also, it is expected that such events would occur relatively infrequently, and therefore Markov analysis should remain valid [9].

2.5.2   Reliability Modelling Tool

The reliability modelling tool used by BT has been developed over the past decade. It originated in 1993, and was then based on an empirically derived algorithm obtained from Monte Carlo analyses of $1 + 1$ and $N + 1$ redundancy studies. The Monte Carlo technique is a statistical simulation of the physical system or process, where behaviour is described by probability density functions (PDFs) that are chosen to closely resemble the real system. A simulation proceeds by randomly sampling from the PDFs, the desired result being an average of multiple observations performed over time.

By 1995, the model had evolved to include calculations for protected paths through networks comprising equipment and infrastructure sections. However, the Monte Carlo approach was limited in terms of accuracy and the range of failure rates that it could accept. These limitations were addressed in 1997 with the first production of the current Markov-based version of the model.

The availability tool is capable of representing many aspects of 'real-world' reliability that are often overlooked or assumed to be negligible in some models. These factors include the following.

- Imperfect fault coverage

  Fault coverage is the probability that any protection/restoration method is successful and is often assumed to be equal to 1. This is clearly not correct as there is a finite chance that any such process will fail.

- Dependent failures

  These can be:

  — either system impairing, where a fault on one component impairs performance of another, e.g through temperature variations;

  — or common cause, where a single event causes multiple faults, e.g. the power supply to a multi-unit shelf fails.

- Latent (or hidden) failures

  This is where a fault remains undetected until a failure occurs that requires the use of that component/path. An example would be a protection path that has suffered a break that is not noticed until that path is required to recover another failure.

- Repair-induced failures

  Faults caused while another problem is repaired, e.g. (accidentally) removing another working component when replacing a faulty one.

It is also possible to define separately the FRU repair times for service and non-service affecting failures, reflecting how a network operator would prioritise certain repair tasks over others.

The tool can model both equipment, in terms of FRUs, and infrastructure, such as fibre, buildings, power, internal ties, etc. In particular, the fibre is sub-categorised into intrinsic faults, namely those due to individual fibre failures, and extrinsic faults, from damage to entire cables/ducts. Such incidents can of course be due to the operators themselves, contractors working on behalf of the operator, or unrelated third parties. Using field-measured fault rates and repair times from various BT platforms and networks and predicted data from equipment and infrastructure suppliers, it has been possible to construct a large database of components. This has allowed BT to extensively formulate product quality of service (QoS) guarantee levels and check the effect on end-to-end services of various equipment, architectural and strategic network modifications.

2.5.3   Protection and Restoration Path Availability

Consider a circuit between nodes 35 and 32 in the generic network model (Fig 2.1). In the 1+1 dedicated protection case, the back-up path is as shown in Fig 2.8(a); the path is known before any failure event and is solely for the use of that particular circuit. Its availability is simple to calculate using the BT reliability tool, and would be of the order of 99.99x% (where x depends on the actual equipment deployed).

The preplanned path shared restoration back-up route, shown in Fig 2.8(b), is also known before a failure, but will most likely have a slower switch-over time than the dedicated protection mechanism — a few hundred rather than a few tens of milliseconds. This does not have as significant an impact on the end-to-end availability as one might think because the reliability of any circuit is dominated by the fibre/duct failure rates and repair times (which can be as high as tens of hours for major cable hits).

The more significant factor, and the one where the complications truly arise, is that the capacity on the subspans used by the back-up path can be shared with other restoration paths, and if any section of the back-up path is unavailable, the restoration will fail.

It is theoretically possible to estimate the probability that capacity on any subspan will really be 'spare' when an incident occurs. However, to calculate this uncertainty you need to know what other circuits share that restoration route, how much of the capacity they would require during a failure, how often they would want to use it and are those other failures statistically connected, e.g. do they always occur together or are they totally random.

For the dynamic path and adjacent span restoration schemes, the calculations become even more complex because the route of the back-up path is determined by the location of the failure on the working path. Obviously, this means there are a range of possible restoration routes available to any circuit.

Although some of the above data is available from SARC, a suitable extension to the reliability tool has not yet been completed to model this type of circuit. Work is under way to develop software that is capable of considering multiple and sequential failure scenarios, once again building from the fundamental state analysis principles of Markov modelling, which should aid in solving these more complicated problems.

2.6   Summary

Near-future multimedia communications will require high-capacity networks employing optical WDM technology to simultaneously carry increasingly large numbers of broadband users. Fast and efficient restoration of network service after failure will be critical to maintaining customer satisfaction and ensuring business success.

This chapter has discussed state-of-the-art software tools and algorithms developed by BT for automated design of such networks, planning of their restoration/resilience capacity and calculating end-to-end service availability.

As the tools and underlying algorithms/models are largely technology-independent, they may be applied to a wide variety of scenarios. In this work, a generic network model representing an inter-city network for a large European country was considered as a realistic test case.

With regard to network design, the comparison between mesh and ring-based architectures was investigated, subject to the (often-conflicting) constraints imposed by the available duct network and customer requirements.

A very large number of different candidate topologies were explored. For the particular traffic scenario under consideration, the following conclusions may be drawn:

- the BT mesh design algorithm successfully reduced the number of links required by 15%;

- mesh capacity was utilised in an efficient manner, with strong correlation between link and switch behaviours, as expected;

- the BT ring design algorithm identified twenty-four rings as best serving the given traffic demands, selected from an initial pool of several hundred candidate rings;

- the profile of traffic load across each ring was much less uniform than the mesh case, emphasising the dominance of a relatively small number of rings.

The above mesh network design was then audited and optimised for various shared restoration schemes, demonstrating the capabilities of the BT developed SARC tool. This showed the potential reduction in spare capacity for that network, where:

- 1+1 dedicated protection used 160% extra spare capacity;

- preplanned path restoration used 108% spare capacity, reducing to 100% with optimisation;

- adjacent span used 140% spare capacity;

- dynamic path restoration used 91% spare capacity.

It also illustrated the impacts and issues that have to be considered with each type of restoration, including the variable length of back-up paths and the speed of recovery from failure.

Finally, BT's reliability modelling tool was introduced, and the complexities associated with calculating the availability for a circuit using shared restoration were discussed.

References

1   Hayes, E.: '*Operators and users consider network resilience issues*', Telecom Markets, Issue 440 (November 2002).

2   Kershenbaum, A.: '*Telecommunications Network Design Algorithms*', McGraw-Hill (1993).

3   Johnson, D. and Hayman, N.: '*Carrier network architectures*', White Paper, Issue 1.1 (July 2002).

4   Arakawa, S., Murata, M. and Miyahara, H.: '*Design methods of multi-layer survivability in IP over WDM networks*', Proceedings of SPIE, **4233**, Opticomm (2000).

5   Fumagalli, A. and Valcarenghi, L.: '*IP restoration versus WDM protection: is there an optimal choice?*', IEEE Network (December 2000).

6   Schallenburg, J.: '*Is 50 ms restoration necessary?*', IEEE Bandwidth Management Workshop (July 2001).

7   Billinton, R. and Allen, R. N.: '*Reliability Evaluation of Engineering Systems*', (2nd Edition) Kluwer/Plenum (1992).

8   Lewis, E. E.: '*Introduction to Reliability Engineering*', (2nd Edition) Wiley (1996).

9   Ford, D. J.: '*Markov analysis of telecommunications systems having imperfect fault coverage*', MSc Dissertation, City University, London (July 1995).

3

STRATEGIC NETWORK TOPOLOGY AND A CAPACITY PLANNING TOOL-KIT FOR CORE TRANSMISSION SYSTEMS

C D O'Shea

3.1 Introduction

Let us first set the scene.

Core transport network design and capacity planning tasks have become increasingly complex over the recent past due to a number of factors. These include:

- the explosive growth in broadband product offerings, driving the migration from narrowband to broadband transport network architectures;

- the range of technologies that can be applied in the core transport arena, including SDH point-to-point, shared protection rings and optical WDM — in fact, the BT network consists of all these technologies as well as the legacy PDH network;

- the scale of the network — the BT core network infrastructure consists of over 10 000 switches and 20 000 connections, which makes manual planning processes virtually impossible.

To make the most effective use of the available capacity and to support viable business cases for future build plans, advanced tools and techniques are an absolute necessity. This chapter discusses:

- high-level detail of the routing and optimisation algorithms used within the tool-kit as well as describing some of the data modelling and scalability issues;

- the delivery mechanisms used to make the tool-kit available to the user community;

- the business benefits of using the tool-kit and how it plays a decision-support role in key areas of strategic network planning, citing a number of case studies.

3.2   Strategic Capacity Planning

What is the tool? The strategic capacity planning tool-kit, known as C-Plan, is primarily designed to assist network planners in assessing the impact of a bulk forecast demand on the transport network in terms of the additional capacity/ equipment required to support the demand. In addition, there are components of retrospective design contained within the solution.

C-Plan does not model the network down to 'nuts and bolts' level. It does, however, model the network in enough detail to determine when capacity on key network elements will reach exhaustion.

In order to run the tool-kit the following data sets are needed:

- the logical network infrastructure (e.g. connectivity);

- the physical network infrastructure at the duct level (for physical diversity);

- a set of switches;

- a set of demands to be routed over the network (forecast);

- the product (or service) type for the demand;

- a set of rules that relate to the physical design and routing protocol;

- the existing 'frozen-in' capacity.

Utilising the above data will enable C-Plan to identify the following:

- where there are likely capacity bottle-necks;

- the incremental capacity requirements;

- where the addition of new switches and connectivity will improve the overall routing efficiency;

- where changes to the routing rules can improve the overall routing efficiency.

It is possible to predict the load on network elements for a given traffic forecast. The traffic forecast is broken down into a number of product lines and an output report provides details of the load on each network element by product type. The view of network load by product type is important as network planners are often interested in knowing which product is driving capacity to exhaustion in certain parts of the network. The existing or 'frozen-in' traffic on the network is also downloaded from the operational database. These are circuits that are carrying live traffic and are not intended for re-routing.

3.2.1   Data Model

It should be noted that the data model used by the tool-kit is an abstraction of the data provided by the operational system. This is necessary as the tool-kit is strategic/ tactical and need not concern itself with every single network element. The definition of the data model is of paramount importance when building/evolving the system. There is a compromise between modelling too much detail versus building a representative real-world network model.

An important requirement while building the data model is to include details of in-station connectivity. In large transport-based networks where there are multiple switches in the same locality, the in-station connect is a key element in planning and dimensioning the network. It is not possible to plan in-station connectivity in isolation from the rest of the network as the connectivity will be determined to a large extent by the traffic flows from the external connections through the building.

C-Plan has a generic model for in-station connectivity to allow corresponding traffic flows to be analysed. In-station modelling becomes more of an issue when there are multiple switches in the same locality.

3.2.2   Forecast Engine

It is often said that 'a tool is as good as the information supplied to it'. Generating an accurate traffic forecast for forward capacity planning must be an overriding consideration.

The forecast input for the network is based upon two data feeds:

- operational database trend;
- product line.

There is a facility to extract circuit-routing information from the operational database (usually the last 6 months) and project this forward in time, based upon comparison of successive financial years of the volume forecast growth. This process is required in the absence of detailed circuit forecasts. C-Plan is designed to interface with the operational platform directly, thereby allowing an accurate extract of the circuit data to be obtained. This is then manipulated in what is known as the 'Demand Wizard' part of the tool-kit to complete the forecast construction.

Alternatively, if a forecast is available, then this can be imported into C-Plan directly.

The forecast usually extends over a 2-year period, allowing investment teams to plan adequate network infrastructure.

3.2.3   Data Capture and Visualisation

The capacity planning functions within the tool-kit have an interface to the operational planning and inventory system through a client/server interface.

Figure 3.1 shows the high-level system architecture for the C-Plan client/server application. The components of the system are described here in terms of a three-tier architecture.

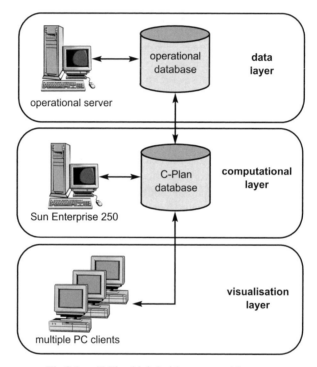

Fig 3.1   C-Plan high-level system architecture.

The data layer for the C-Plan system is provided by an external database known as the planning, assignment and configuration system (PACS). The database holds information describing the target network and is read-only. The data is downloaded to the computation layer prior to any manipulation.

The C-Plan server provides the computation layer for the C-Plan system. It is a Unix-based system, supporting an Oracle database. The server holds the input data preparation tools, the C-Plan routing and optimisation tools, and the output data preparation tools. A request server controls the execution of these tools and processes user commands delivered via the visualisation layer. The C-Plan routing engine and optimisation routines are a separate entity encapsulated within the C-Plan framework. The size of networks catered for depends on the complexity of the problem and the algorithm used. However as a benchmark, C-Plan has been deployed to capacity manage the SDH transport network for BT. The data model for this network consists of over 10 000 switches and 20 000 connections. The run time to provide a capacity plan is typically 30 minutes maximum on a Sun Enterprise 250 Computer (C-Plan will also run as a stand-alone package on a desktop PC with a standard build).

The visualisation layer for C-Plan is provided by a Visual Basic client application. The application allows multiple users to use the C-Plan system from their desktop PC. The client application provides the users with the functionality to define input sets, kick off processes on the server and present the results of the C-Plan runs. It is possible via the client to cap existing infrastructure, i.e. to use a structure to its bandwidth limit but not to deploy more of the same. It is also possible to add new structures (based on a set of planning rules) in order to carry out 'what-if' reactive design (see section 3.2.5).

The output files can be downloaded into Excel, Microsoft Access or an Oracle database as the requirements dictate. The files can also be presented as MapInfo[1] tables, and a utility is provided to enable the network and output to be viewed graphically.

Figure 3.2 shows the MapInfo[1] interface with a large network. Thematic mapping is used to colour code the connections according to their load and to identify localised capacity bottle-necks.

3.2.4   Route Optimisation Platform

There is likely to be a wide range of routing strategies and specific problems to investigate with respect to achieving a minimum cost network. It therefore makes sense to build a routing engine that is generic and can be tailored to a particular problem without a large overhead in re-development costs. Routing rules can be applied to the use of least-hop or least-distance configurations or some other cost metric. The routing software is also able to deal with networks of a hierarchical nature with associated routing rules for transiting between layers.

Static routing protocols can be considered (such as simple shortest path routing) as well as 'dynamic network-state' routing — where the route depends on the current state of the network. Various routing metrics can be defined to try alternative routing strategies and analyse the effect of this on the network and associated costs.

The routing engine can route circuits either via a single route or via a primary and back-up route that are physically node and link disjoint (down to the physical layer, if required, i.e. ducts). Hence, it is possible to analyse the cost of providing this type of network protection as opposed to using alternative schemes such as restoration [2].

---

[1]MapInfo [1] is a commercial off-the-shelf graphical information system product which has been integrated with C-Plan.

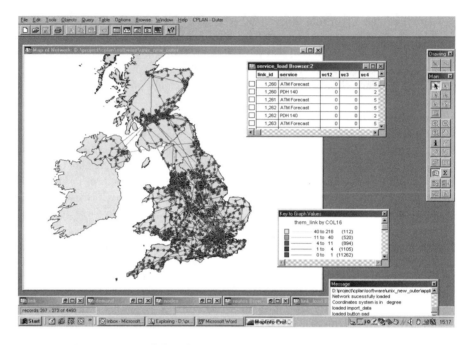

Fig 3.2   MapInfo interface to C-plan showing network infrastructure.

Specialised routing software is also included if there is a requirement to minimise whole network costs via an intelligent routing algorithm rather than routing each demand by shortest path. The 'maximum flow' algorithm can be used, for example, to increase the packing efficiency when grooming VC-12s on to VC-4 highways within the SDH network layer or for determining an optimal set of wavelength paths in an optical dense wavelength division multiplex (DWDM) network layer.

The routing efficiency can be influenced by the choice of an in-station termination point. For example, for a circuit to route from Edinburgh to Watford as shown in Fig 3.3, there are eight possible route termination combinations (e.g. Edinburgh:SDX via Newcastle to Watford:ADM2 via Kingston). The routing engine will choose the least-cost route and termination points for the route. In the case of route diversity, the least-cost set of diverse routes and termination points will be chosen. Section 3.3.3 provides results of a case study on the benefits of termination ends optimisation.

3.2.5   Structure Planning

There are two basic mechanisms for performing structure planning. These are:

- retrospective design;
- pre-emptive design.

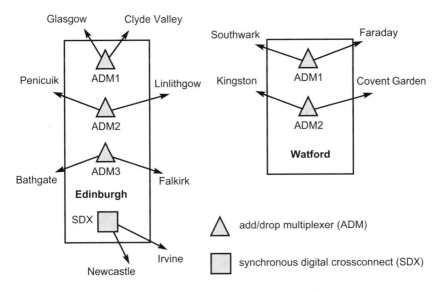

Fig 3.3   Example of termination ends problem.

Retrospective design involves:

- running the dimensioning module C-Plan, within the tool-kit;

- examining the resultant network load distribution;

- altering the network retrospectively by capping or creating new structures manually within the structure wizard (available within the C-Plan client) to improve the load distribution.

Pre-emptive design involves:

- designing a network based on a physical connectivity plan (i.e. ducts);

- a set of candidate sites;

- a traffic forecast.

In this pre-emptive case the design process is moving towards using computing power with algorithmic techniques to propose a structure design.

The choice of design depends on a number of factors, the main one being whether there is a step change in deployment of new infrastructure or if the network is catering for natural growth. Pre-emptive design is discussed in more detail in Chapter 2.

Section 3.3.3 describes a scenario where the resulting load on the network has been used to determine which ring structures are likely to have poor fill. This may be because there is a dominant switch on the ring that is causing a load imbalance. This will be the 'trigger' to perform retrospective design analysis to find a solution.

3.3   C-Plan in Action

The following scenarios describe the application of C-Plan in a strategic support role.

3.3.1   Scenario 1 — SDXC Port Capacity Exhaustion

3.3.1.1   Problem Statement

The problem can be described with the aid of the diagram in Fig 3.4. The diagram shows high-level in-station connect between add/drop multiplexers and the crossconnect. The crossconnect has a maximum capacity of 504 STM-1 I/O ports. If we consider an individual STM-1 connection between ADM1 and the crossconnect, the STM-1 may only contain one VC-12 (worst case). The total capital cost of the crossconnect is of the order of £1.5M. Hence, each port will cost the equivalent of £1.5M/504, i.e. £3k. If there are an appreciable number of this type of connection with low fill, then the crossconnect port capacity could reach exhaustion very quickly even though the switch matrix may be under-utilised. In addition, the traffic may want to flow from one add-drop multiplexer to another in-station, without even needing to use the crossconnect at all.

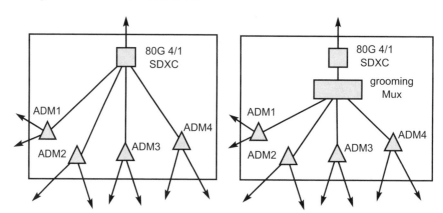

Fig 3.4   Insertion of grooming multiplexer in front of crossconnect.

To address the problem a 'grooming multiplexer' needed to be deployed before the crossconnect in order to groom the circuits on to connections before they reached the crossconnect, thus saving on port capacity. However, it is not a simple matter to determine a priority roll-out list of where and when the 'grooming multiplexer' should be deployed as this will depend largely on the traffic flows in-station.

3.3.1.2   C-Plan Impact Analysis

By using the C-Plan in-station data model, the grooming multiplexer functionality was modelled in every locality where a crossconnect was deployed.

It was then possible to determine which grooming multiplexers took the greatest load based on a two-year forecast and thus determine a priority roll-out list to key sites.

3.3.2   Scenario 2 — Wavelength Configuration Plan

3.3.2.1   Problem Statement

For optical DWDM-based networks, wavelength paths are set up between localities based upon the predicted traffic flow. Wavelength paths can bypass (or 'glass-through') localities if there is insufficient add/drop traffic. The problem is to configure an optimal set of wavelength paths at either 2.5 Gbit/s or 10 Gbit/s across the network to support the traffic requirements. The majority of this traffic will be high-order SDH bearers at STM-1 and above.

3.3.2.2   C-Plan Impact Analysis

C-Plan has an embedded optimisation algorithm known as the 'maximum flow' algorithm. The algorithm will identify the wavelength paths in the network based on a set of planning rules and a traffic forecast.

Figure 3.5 shows a graphical representation of a set of 600 point-to-point demands corresponding to 400 Gbit/s of high-order SDH bearer circuits to be mapped on to an optical network.

These demands are 'groomed' into wavelength paths as defined by the optimisation routine to minimise the number of wavelengths provided with low fill factor (fill factor is the number of VC-4s per wavelength).

In the worst-case scenario a wavelength could be set up for each unique demand in the network.

Since there are 600 unique demand pairs, this would result in 600 wavelength paths or 690 distinct wavelengths at 2.5 Gbit/s being configured. Figure 3.6 shows the wavelength load distribution for the demand set after grooming, comprising 200 wavelength paths or 350 distinct wavelengths at 2.5 Gbit/s, representing a reduction of 49%.

As can be seen, there are a relatively small number of wavelengths (between 3 and 4) that have a fill factor of less than 10 VC-4s.

Fig 3.5   Representation of demand set for wavelength configuration plan.

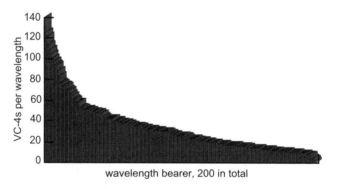

Fig 3.6   Wavelength load distribution to support the demand.

3.3.3   Scenario 3 — Termination Ends

3.3.3.1   Problem Statement

As discussed in section 3.2.4, when there is a choice of equipment termination in-station, the decision can affect the routing efficiency. It was necessary to quantify the impact of this in order to determine if it was worth the effort of employing additional support software within the operational platform to provide a better means of administering this task.

3.3.3.2   C-Plan Impact Analysis

To study this in a large transport network, 5000 VC-4 (155 Mbit/s equivalent) circuits have been routed on the broadband network using C-Plan under two different scenarios. The first scenario is where the termination ends have been chosen at random; the second scenario is where the routing engine determines the 'least-cost' termination ends to use.

The total number of VC-4 circuits was summed over all the broadband connections in the network and the results are as follows:

- for random choice of termination ends, total number of VC-4s = 17 452;

- for the routing engine choice, total number of VC-4s = 13 944.

This corresponds to a significant 20% saving in network capacity due to improved routing efficiency.

3.3.4   Scenario 4 — Ring Load Balance Analysis

3.3.4.1   Problem Statement

With ring-based network architectures, one of the challenges is to design a set of rings with good load balance capability. Ring structures require the switches and connections on the ring to support the same level of bandwidth. For example, within the SDH layer, rings can be supplied at bandwidth rates supporting STM-1, STM-4, STM-16 and even STM-64. If the traffic load across the multiplex sections (and switches) on a ring is not evenly distributed, then bandwidth (and equipment) can be wasted. A hypothetical example of this is as shown in Fig 3.7 where ADM1 is a 'dominant' switch on the ring and traffic sourced from this switch is trying to leave the ring via gateways ADM2 and ADM3 to reach an off-ring destination (in the example, traffic is designated 1+1 protected in SDH and requires both a primary and a back-up route). The other switches, ADM4, ADM5 (being relatively small), do not

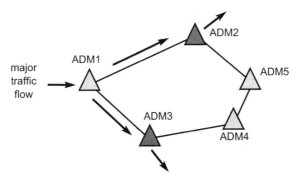

Fig 3.7   Example of 'poor' ring load balance.

pass any appreciable traffic across the ring and as such the load balance on the ring is poor.

3.3.4.2   C-Plan Impact Analysis

At present, we have used qualitative measures of how well a ring is load balanced employing terms such as 'good' and 'poor'. It would be far better to use a more precise way of assessing the load balance on a ring structure. Within C-Plan a quantitative measure of load balance has been defined and agreed with the network planners. This is known as the traffic load 'disparity' and is derived through the relative difference between the mean traffic on the ring (across all multiplex sections) and the standard deviation of traffic on the ring.

C-Plan can be used to provide a structure report that gives details of the traffic load disparity for each ring structure in the network. Figure 3.8 illustrates this for a network based on shared protection rings serving a broadband demand traffic forecast. Armed with this data, a network planner can set a threshold above which action will need to be taken and this may involve reconfiguration. This approach provides a methodology for highlighting problem areas in the network in order to perform elements of retrospective design.

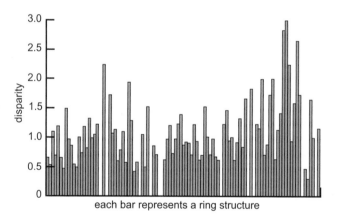

Fig 3.8   The three types of V-structure.

3.4   Summary

The application of routing and optimisation algorithms encapsulated within the C-Plan tool-kit has been highlighted. A number of scenarios demonstrate that tangible benefits can be achieved by the application of the tool-kit in the area of strategic network capacity planning.

A key element in the success of C-Plan and its continued use has been the definition of a data model that is sufficiently flexible and scalable to handle the vast majority of the problems that are raised by network planners.

The ideas and principles behind the C-Plan tool-kit are essentially generic and can be applied to a range of technologies and network architectures.

References

1   MapInfo — http://www.mapinfo.co.uk/

2   Restoration — http://www.cstp.umkc.edu/jnsm/vols/vol02_abs/02-2-doverspike.html

4

A BAYESIAN NETWORK DATAMINING APPROACH FOR MODELLING THE PHYSICAL CONDITION OF COPPER ACCESS NETWORKS

D Yearling and D J Hand

4.1 Introduction and Review of Previous Attempts

This chapter looks at how the condition of existing access networks can be modelled.

The PSTN copper access network forms a major transport channel for successful delivery of broadband services to residential customers in Britain [1]. Using asymmetric digital subscriber line (ADSL) technology, the twisted copper pair used for telephony can also carry digital data traffic by passing higher frequencies down the line. In theory, a copper pair has well-defined characteristics for signals up to several megahertz. However, BT has a PSTN network that was installed primarily for low-frequency voice traffic, some of which has been in service for over 30 years. Clearly, it is impossible to know the physical characteristics of all copper pairs but this information along with their age and condition have a real effect on high-bandwidth traffic (in terms of attenuation, crosstalk and noise). Consequently, within an aging network this can result in an increased fault rate and poor customer satisfaction.

For many reasons it is sensible to assume that the fixed line network will still be the transport of choice for some time to come. With the promise of the data wave sweeping the nation, people will inevitably become more data hungry. Despite the great advances in technology BT has made in the development of digital techniques, optical devices and transport, reliance is still made on the copper access network which represents a considerable proportion of BT's fixed assets. Although this technology is somewhat aged, it remains a cost-effective solution to modern data and voice transport requirements.

With this network, BT has a UK coverage which far outstrips that of any other operator. In terms of scale, this represents a service to approximately 20 million residential and 6 million business customers. Therefore, increasing our ability to identify weak or prone plant is of obvious benefit, but, as we gradually take this forward further, improvements become even more difficult to make.

Generally speaking, the relationships between faults and their respective causes are not straightforward. Concerted efforts such as the Access Network Improvement Programme (ANIP) involves extensive electrical measurements of telephone lines during overnight tests.

By using a system of heuristics, nodes in the network are identified as hot spots, sometimes prior to them being reported faulty. Although a technically good solution, in reality the experience has been that initially very poor plant is quickly identified, but subsequent detection of poor performers fails despite a continued elevated fault volume. This principally lies with the heuristics used, but also with the variation and temporal vagaries of the electrical readings themselves.

Weather, materials and even the mode of repair work all have an impact on failure numbers. Quite often engineers have to repair faults in bad weather conditions or on awkward plant. This can have the effect of causing more faults, and these are generally termed intervention faults. The type and health of the network is also important — not only the location (whether the lines are predominantly overhead or underground), but also the technology and materials used. For example, Fig 4.1 clearly shows an aerial lead cable in poor repair which was located in the

Fig 4.1   Overhead lead cable breaking free from its support.

study data referred to in this chapter. In Fig 4.2, a clear view of the age of some of the plant is apparent, this pole has bare open copper pairs alongside more modern insulated services.

Fig 4.2   Very old pole top showing bare copper pair.

A materials issue is the use of aluminium, which was used in the copper access network in the 1970s during a period of high market copper prices. It is far less ductile than copper, and has a larger coefficient of expansion (which is almost twice that of iron). This may present problems in Fe/Cu joints, which experience large temperature differentials. Intuitively, this should only really become manifest where the plant itself is subject to large differentials due to solar gain. In black body experiments, within-plant temperatures can reach 70 °C. Underground, temperatures remain relatively stable. For aluminium to copper joints, the greatest problem is not temperature differentials but moisture. When there is moisture ingress into the joint environment, invariably the water is contaminated, and this allows the joint to become a virtual galvanic cell with the two metals forming the anode and cathode within the electrolyte. Combined with the effect of oxidation, this results in a rapid corrosion of the aluminium (the anode).

With respect to difficult equipment, Fig 4.3 clearly shows an untidy cabinet which will obviously cause problems when an engineer is faced with a diagnostic task and disturbs neighbouring wires. This cabinet also contained aluminium tails, again making repair difficult as mentioned above; add to this inclement weather, and it is far from surprising we can encounter intervention problems. Even when new plant is installed there can be avoidable problems — Fig 4.4 shows a brand new cabinet with supermarket carrier bags as duct seals; these will obviously degrade and provide a poor moisture barrier within a short period of time.

Fig 4.3   Untidy cabinet which also contained aluminium tails.

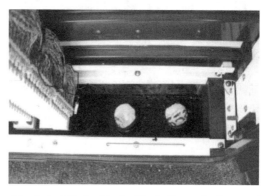

Fig 4.4   Newly installed cabinet with plastic carrier bags as duct seals.

The purpose of this chapter is to propose a Bayesian datamining approach to investigating plant fault propagation. Here we consider all wire and joints from customer premises to the exchanges (see Fig 4.5). As we are attempting to build a causal model, it is important that it is sophisticated enough to bring together a variety of data sources at a fine level of granularity — not only fault records, but weather and geographical information. At the same time, it must be computationally efficient so as to provide scaling when operating over large areas of the PSTN network.

Fault analysis within the PSTN network is not new [1, 2]. Despite this, it is certainly the case that there is still a great deal of improvement to be made on what has gone before. For instance the analysis in Thomas et al [2] is somewhat simplistic, and in some places somewhat obscure and even misleading (see, for example, Fig 6 of that paper in particular). Certainly simple regression techniques

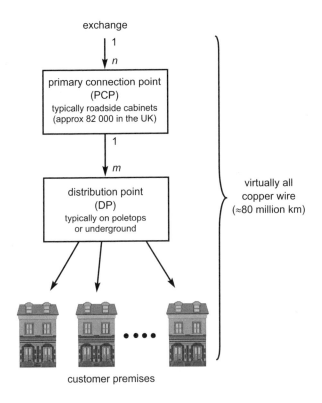

exchange

primary connection point
(PCP)
typically roadside cabinets
(approx 82 000 in the UK)

distribution point
(DP)
typically on poletops
or underground

virtually all
copper wire
(≈80 million km)

customer premises

Fig 4.5 Schematic description of the network serving the typical customer.

are far from adequate for a detailed picture of fault propagation, since the assumptions required are unlikely to be met. Further, this has a direct consequence on any confidence or prediction interval generated. Another aspect is the reported association between fault volumes, manpower and climactic conditions. These relationships are tenuous, based on the data and level of granularity employed. No statistical information is included for the reader to understand the suitability of the modelling done. Certainly the model presented requires this. Manpower (as presented in the paper) is likely to be highly colinear with many other indicator variables that drive fault volumes. It is also worth noting that manpower may be considered confounded with plant age and condition (as explained in Yearling and Hand [3]) — consequently one should be cautious about making conclusions that affect repair processes using this model as confirmatory evidence. A further suggestion that caution should be taken is the findings that half of all faults are caused by intervention. This is not immediately clear from the modelling evidence presented [3]. Furthermore, it is highly unlikely that monthly rainfall figures taken over large areas is going to be a useful variable in any regression analysis. This is due (in part) to the localised effects of rainfall. The approach taken also employs a preconceived model of fault aetiology — we are keen to avoid such a prescriptive approach here.

Taking the previous analysis [3], we encounter a different problem — namely, the identification of a manifold problem space that certainly lies outside the scope of a single review. It is our intention to draw together several strands of previous work in order to present a more cohesive and defensible model. This should take into account many of the aetiological factors which may induce faults and also explain their respective relationships, interactions and impact on fault volumes. Another principal objective is to render any resultant models easy to understand and useful for strategic and financial planning within the copper access network.

4.2   Taking a Bayesian Viewpoint

Broadly speaking, there are two schools of thought within the field of statistics — Bayesian and frequentist. Although the differences between the two are philosophical, they have profound implications in data analysis. For many years a great deal of intellectual debate has centred on the relative rights and wrongs of both, and it is totally beyond the scope of this chapter to continue that debate. Instead, we shall outline the fundamental reason why being a Bayesian can be useful. The fundamental difference between frequentist and Bayesian statistics is the way in which probabilities are formed and used.

In a frequentist world, we have absolute values for distributional parameters. For example, consider the distribution of working lifetimes of a given telecommunication switch. For the frequentist the parameters of that distribution are fixed (but perhaps unknown) in value. In our lifetime analogy, if we consider the parameter of interest as the mean lifespan, it has a constant value and any observations (random variables) taken from the population of all possible switch lifetimes would be consistent with variation about that mean. Unfortunately, although this is mathematically convenient it holds problems when we come to estimate our mean lifespan from a given sample. We are totally unable to form any probability measure of how far (or near) our sample mean is from the true mean. For example, 'confidence intervals' convey no probabilistic information about our sample mean whatsoever.

In a Bayesian context, this makes no sense at all. In the Bayesian world, nothing is fixed — nothing is constant, irrespective of whether it is the mean switch lifetime or even a physical constant such as the speed of light. Here, our sample of switch lifetimes is truth, reality, and our perception of the world. Our population parameters are the real random variables. In this example, the mean switch lifetime is a random variable that has an associated probability distribution. Using this we can form probabilistic statements about our sample mean. A corollary of this is that a Bayesian distribution of interest (q say) is conditional on any information (I) and data/observations (y) we may have. This is achieved by using the ubiquitous Bayes Theorem:

$$P(\theta|y, I) = \frac{P(y|\theta, I)P(\theta|I)}{P(y|I)} \qquad\qquad ......\ (4.1)$$

Thus, we can update from our prior distribution for θ (namely $P(\theta|I)$) to our posterior distribution $P(\theta|y, I)$ given any new data y.

This fundamental difference in viewpoint is often the source of many statistical misunderstandings. People frequently have a Bayesian view when dealing with their common everyday experience, but then may apply frequentist analysis. The common misinterpretation of confidence intervals is an instance of this. Furthermore, most people have a prior conception as to the likelihood of events that affect their lives, even if they have never experienced the event. For example, many people constantly evaluate personal risks when driving, taking a flight, or eating particular foods. People also readily accept that these events are often conditional on the outcome of other events (these may be known or unknown as the case may be). Graphical models allow us to tentatively explore this causal world, and by concentrating on a particular subset we can also incorporate a Bayesian probabilistic approach.

4.2.1   Graphical Models and Bayesian Networks

Graphical modelling is a general multivariate framework that brings together various statistical models under a graph-theory notation [4-8]. As such, many traditional multivariate statistical techniques can be viewed as special instances of the general graphical model (examples include structural equation models (SEMs) and hidden Markov models (HMMs)). We shall be focusing on the technique known as Bayesian network (BN) models, which for our purposes has distinct advantages.

4.2.2   Basics of Bayesian Networks

BNs are an effective way of representing a closed system of elements which interact in some stochastic fashion. Like all graphical models, they use graph theory in order to formulate the problem space. For the sake of brevity, we assume the reader is familiar with basic graph terminology and probability and refer to Lauritzen [9], Whittaker [10] and Edwards [11] for further reference.

Consider a directed graph, $(N,\vec{A})$, consisting of the node set, N, and the set of directed arcs connecting these nodes, $\vec{A}$. Each $n \in N$ corresponds to a system variable, and each $e_k \in \vec{A}$ represents the conditional relationship between any two nodes x and y in N. Following standard probability theory, if RVs X and Y are mutually independent, then:

$$f_{XY}(x, y) = f_X(x)f_Y(y) \qquad\qquad \text{...... (4.2)}$$

The idea of conditional independence is an extension of this, where now we consider three RVs (say X, Y and Z).

Here, if X and Y are independent given the realisation $Z = z$ (that is, information relating to the true state of Z), then $X \perp\!\!\!\perp Y|Z$ (X and Y are conditionally independent given Z). Probabilistically speaking (following equation (4.2)) we have the definition as given in equation (4.3):

$$X \!\perp\!\!\!\perp\ (Y \Leftrightarrow P(x|y, z) = P(x|z)) \qquad\qquad \text{...... (4.3)}$$

In terms of a graphical representation, the conditional independence $X \perp\!\!\!\perp Y|Z$ takes the form as shown in Fig 4.6 with the three nodes representing the variables and the absence of arcs implying conditional independence. Following the notation, the arcs $[ZY]$ and $[ZX]$ imply that $Z \not\!\perp\!\!\!\perp X$ and $Z \not\!\perp\!\!\!\perp Y$. Conditional independence is a fundamental concept needed for the creation of the networks described in the following section.

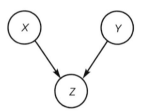

Fig 4.6   Conditional independence.

From Fig 4.6, it is also worth noting that in this instance the node Z is defined as a collider with respect to the arcs from X and Y, i.e. they directionally converge on Z. If this were not the case (for example, if the direction followed through Z from X to Y) then conversely Z would be a non-collider. This condition relates solely to the adjacency path concerning the node in question, and is not a property of the node itself. Thus, any node can be a collider and non-collider simultaneously with respect to different adjacency paths in the network.

By directing the links, we represent the concept of causal conditioning. The strength of any causal link is given by attaching a conditional probability table to every node, where each entry represents a conditional probability statement regarding the states of the node in question in relation to all possible states of its parent nodes. In order to facilitate this we make each node discrete, allowing it to take any number of finite states. We avoid intractable probability calculations, by enforcing that no feedback loops are contained within the graph.

4.3   Forming the Problem Space and Data Description

The data and problem space follows that previously defined [2], and will only be briefly summarised here. The focal point of the study was chosen to incorporate a high number of faults (see Fig 4.7). This is important as, at the network node level, faults are rare indeed. This adds a great deal of complexity in terms of modelling in the statistical sense. Therefore, it is important than an inner city area is chosen (with corresponding high network density), coincidental with a high-quality point source of weather data. The focal point of the study (the University of Plymouth (see Fig 4.7)) has an automated weather station providing a detailed view of the weather in the city centre in ten minute intervals. This city centre location is fortuitous, as most weather stations are located in remote areas (such as air fields) where network density is low. The exchange serving the majority of customers within a 1 mile radius of this focal point was determined, and the fault reports were aggregated with respect to the distribution points (DPs) which connect customers to the principal connection points (PCPs). In this study, we define a fault as any maintenance activity that involves accessing the network to rectify a customer-reported problem. We extracted daily fault counts for each DP over a period of one year. There were no missing values, which is vital in order to avoid censoring problems in the data series. This area performs well in terms of fault rate, some 95% of the 1108 DPs experienced less than 4 reported faults for the period. Those with such low fault counts afford little scope for modelling. Therefore, the first filtering process removed these DPs allowing us to focus on the remaining 54. The intention is to model each DP in turn. However, this study will be confined to a single example for the purpose of illustration.

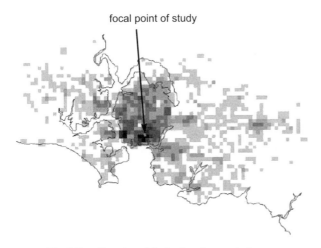

focal point of study

Fig 4.7   Density of faults for chosen study area.

Some temporal clustering is shown by many of the data series in this subset. Typically, they appear to be short bursts of fault activity rarely exceeding a few days. This is indicative of intervention faults — faults caused by engineers entering the network to effect repair but inadvertently causing further defects in the process. This can be a result of poor workmanship, plant or a combination of both. However, we know the vast majority of plant within this area is sound by virtue of an overall low fault count and that all engineers are available to repair all plant. This implies that plant condition rather than poor workmanship may be the principal factor. It is this particular fault pattern that is of interest here. Alongside this, we wish to investigate the additional effect of climate in fault propagation.

Weather variables were collected from the University's automated weather station. All data was thoroughly and painstakingly cleaned. This removed many electronic glitches, instrument drifts and reset anomalies. From the resultant clean set, daily summary figures were calculated in line with the daily fault figures for the DPs. This resulted in eleven weather variables. However, as many of them are likely to be highly colinear, a short pre-modelling analysis identified the most useful using crude Poisson models. For underground faults, humidity was a factor as an illustrative variable for dampness. While for overhead faults, driving easterly wind and rain (alongside high humidity) were identified as prominent. This was later verified with a field audit; engineering practice favours mounting the pole top boxes on the reverse side to the prevailing winds (predominantly westerly), and thus in adverse easterly conditions potential problems are exacerbated. Therefore, the data input into the modelling process were rainfall, humidity and wind speed alongside the fault data series for each DP.

4.4   Learning Causal Structures — Constructing the Bayesian Network

As mentioned earlier, BNs, once constructed, provide an intuitive and insightful model in many applications. This insight gives better understanding and the model output itself can be used directly to make decisions regarding our system under study.

Unfortunately, we are faced with a difficult problem, namely defining the conditional relationships we wish to examine. As we are using the BN as a data exploratory tool, the structure is part of our solution, and so clearly unknown and purely down to conjecture. BNs have been used extensively in situations where such relationships are well known (by experts), and are constructed on this basis, with the probability tables being updated from real data. Examples of this include medical diagnostics and genetic models.

Therefore, we are faced with the task of learning the structure itself. Several methods were tried, but were relatively poor compared with our final choice [12]. It has been shown this award-winning approach will discover the underlying directed acyclic graph (DAG) provided we have sufficient data (and, of course, provided that the data was formed by a BN). This approach is very well described and documented, and so only a brief overview of the salient features apropos this application will be given here.

The technique is focused on conditional independence and specifically relies on dependence separations (d-separations). Any two nodes X and Y may be considered d-separated if all the colliders that lie on the path are elements of a cut-set between X and Y. If not, they are conversely d-connected. This can become conceptually difficult at instantiation (evidence is entered into the network for updating); so simplistically we shall restrict our definition in line with the algorithm as described in Cheng et al [12]. Thus, for the trivial example given in Fig 4.6, X and Y are d-separated. Further, following Cheng, we use the analogy of information flow along the arcs of the network, with the non-colliders and colliders acting like sluice gates (open and shut respectively). Thus, if all information flow between X and Y is blocked by colliders, then X and Y are d-separated.

Crudely speaking, the algorithm works by constructing condition sets, where intermediate sluice gates are open and shut and the flow of information is observed. In this way, we can determine which series of gates block flow and use this to form the structure of the network. The information flow itself is monitored using mutual information (that is, if X and Y are dependent then evidence from one node conveys information regarding the other). In the probabilistic sense, we have the mutual information (and conditional mutual information with respect to any node(s) C) between X and Y as equation (4.4) and equation (4.5) respectively:

$$\text{MIF}_{X_i, X_j} = \sum_{x_i, x_j} P(x_i, x_j) \ln \frac{P(x_i, x_j)}{P(x_i)P(x_j)} \qquad \text{...... (4.4)}$$

$$\text{MIF}_{X_i, X_j | C} = \sum_{x_i, x_j | c} P(x_i, x_j | c) \ln \frac{P(x_i, x_j | c)}{P(x_i | c)P(x_j | c)} \qquad \text{...... (4.5)}$$

How equations (4.4) and (4.5) are calculated is straightforward. Each variable included in the data set is a node in the BN. Both equations are estimated from the relative frequencies contained in the dataset (each record in the study database relates directly to a complete realisation of the network — this is known as an instantiation).

To construct the network itself, we allow provision for constraints (such as event ordering and enforced independence) at each stage of the construction process. However, the extent to which we rely on expert opinion to supply this information is minimal as this learning process may be applied totally unsupervised.

The learning process begins with calculating equation (4.4) for all pairs of nodes (variables). Those pairs with mutual information greater than a given threshold are ranked. Then, beginning with the largest, each pair in turn is visited and joined with an arc provided they are not already d-connected. This results in a draft DAG, which may have some necessary arcs omitted (due to the crude heuristic ranking approach). To do this the algorithm scans the node pair set a second time. At this stage it is attempting to find a minimum cut set (i.e the smallest possible) which will cause the pair in question to be d-separated. We then apply a conditional independence test (using equation (4.5)) to evaluate if the two nodes are independent, given the cut set. If they are not, then an arc is added between the two nodes.

At this point, the algorithm should have found all the true arcs (provided there is sufficient data, of course). Unfortunately, at this point it is also possible (and in fact quite likely for large complex networks) that there is redundancy in the form of too many arcs. This, in part, may be due to the cycling through the node set where essential arcs are yet to be included. There are other rare situations where this may occur, but this approach is robust in such circumstances (indeed, under the same conditions it is difficult for any technique to prosper). To remedy this, the algorithm repeats the testing, but, this time, taking each pair of directly connected nodes, the arc is temporarily removed and tests for d-separation. If the nodes can be separated, then the arc is discarded; otherwise it is reinstated.

Once this is done, the orientation of the graph is then examined. This is certainly the most difficult task to automate, and may result in an inability to orientate certain ambiguous relationships. Orientation is done by using colliders, and V-structures (Fig 4.8). V-structures are simple arrangements of three nodes connected with two directed arcs. However, recall (from Fig 4.6) that only when they are convergent arcs are we able to use conditional independence, and this only occurs in one of the three possible structures. Thus, we begin by detecting all pairs of nodes which are the source nodes for this particular convergent V-structure. From this, it identifies all the colliders between them. This is repeated for all possible V-structures. We then use these identified colliders to orientate the other arcs. This is done in two ways — firstly, if the undirected arc is part of a V-structure (but not of the third kind, obviously) with the other arc directed towards the mid node, we comply with this path and orientate the arc in question away from the mid node; secondly, if there exists an alternative directed path between the source and sink nodes, we comply with this path and direct the arc accordingly. As mentioned above, this is not guaranteed to successfully direct all arcs.

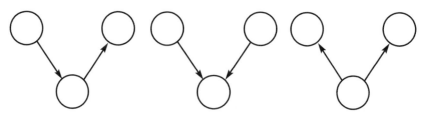

Fig 4.8   The three types of V-structure.

Having briefly described what Bayesian networks are and how the learning process functions, we now turn our attention to the implementation within our problem space.

4.5 Application of Bayesian Network Models for Decision Making

4.5.1 Building the Bayesian Network

We begin this section by specifying both the detail on the data inputs and the required outputs for the modelling process. As mentioned before, it is clear that a highly aggregated modelling approach is not useful and a low level of granularity is essential. Therefore, by modelling at the DP level we have the ability to characterise small network areas and then aggregate them in any way we see fit thereafter. This could be in terms of geography, plant type, age, or placement. All of the weather and fault data sets were discretised using an equal frequency approach prior to the learning process, and detailed in the Appendix to this chapter. We also confine ourselves to simple, intuitive constraints that should appeal to common sense.

With respect to the output, it is important that a small, easily scalable model is produced for each DP. This not only allows the models to be easily understood by engineers, but just as importantly allows them to be scaled in a datamining context. Here, for the purposes of illustration we focus on the modelling of a single DP. However, the computational time required to fit the model illustrated here is certainly in the order of fractions of a second on a standard office PC. This clearly illustrates its potential for learning models over very large areas indeed. The information required from the output should be a set of relationships alongside the strength of such relationships in the form of conditional probability tables.

The modelling was done by authoring a bespoke application (Fig 4.9) written using Microsoft Visual Basic. This enables a graphical interface directly with the model and allows the analyst to interact easily in forming, assessing and exploring the relationships. The application was written purely for ease of use, and indeed using a rapid application language it was equally easy to author. The modelling is done purely visually (Fig 4.9) with variables being dragged and dropped on to the modelling canvas. Once we have placed the nodes in which we are particularly interested, we enter the constraints. Again this was designed to be as straightforward as possible. The only constraints placed on the network prior to learning were the specification of root nodes. These are nodes which may have no parent nodes whatsoever, and are thus unaffected by anything in the model. This type of constraint is perhaps the most easy to defend (e.g. forcing leaf nodes is only really possible with terminal events). We have defined all weather variables as root nodes. This follows intuition, as they are all components of weather and are not specifically

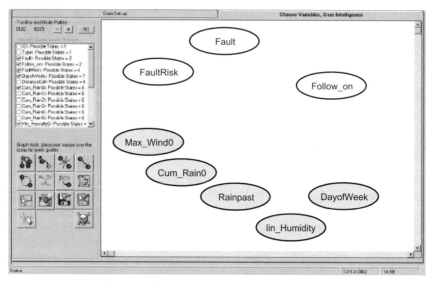

Fig 4.9   Model software prior to model fitting.

caused by one another. Another intuitive root node is that of day of week, indeed it is hard to argue that this node should be anything but a root node. There were no other constraints placed on the network. All of the nodes constrained in the network prior to learning are shown tinted in Fig 4.9 (coloured on the software application).

4.5.2   Explanation of the Bayesian Network

The model construction takes less than a second to run and confirm. The confirmation merely checks that all arcs are directed, and that no loops exist within the network (i.e. the learnt BN conforms to a DAG). The resulting model (for a single overhead DP) is shown in Fig 4.10. Here we can clearly see the relationships learned from the data are consistent with our concerns.

Day of Week clearly has a large influence on whether Fault is true. This is exactly what is expected, as faults directly relate to customer report time (and this follows a clear weekly cycle). Also, only one weather variable impacts on Fault, namely Max Wind. Again, this is what one would expect to find as Max Wind can be thought of as a storm variable which has obvious effects on overhead cables and plant (recall this is an overhead DP). The conditional probability (marginal for day of week) is given in Table 4.1, while Table 4.2 includes the effects of Max Wind. It is worth noting that from the $2 \times 7 \times 7 = 98$ outcomes only those with the highest probabilities when *Fault* = TRUE are given, which also cover all the high winds. It is also worth pointing out that the probabilities remain quite high throughout the week and this implies that these types of fault were reported by the customer very soon after the damage occurred.

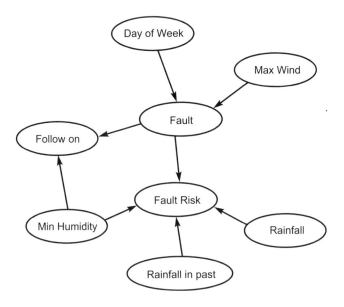

Fig 4.10   The Bayesian network learnt for the study DP.

Table 4.1   Conditional probability of a Fault occurring (marginal for day of week only).

*P(Fault=***TRUE**|*Day=***Monday***)*	0.15
*P(Fault=***TRUE**|*Day=***Tuesday***)*	0.17
*P(Fault=***TRUE**|*Day=***Wednesday***)*	0.11
*P(Fault=***TRUE**|*Day=***Thursday***)*	0.04
*P(Fault=***TRUE**|*Day=***Friday***)*	0.14
*P(Fault=***TRUE**|*Day=***Saturday***)*	0.06
*P(Fault=***TRUE**|*Day=***Sunday***)*	0.07

Table 4.2   Conditional probability of a Fault occurring.

*P(Fault=***TRUE**|Day=**Monday** « *Max Wind=***Greater than 12.45 ms^{-1}***)*	0.56
*P(Fault=***TRUE**|Day=**Tuesday** « *Max Wind=***Greater than 12.45 ms^{-1}***)*	0.50
*P(Fault=***TRUE**|Day=**Wednesday** « *Max Wind=***Greater than 12.45 ms^{-1}***)*	0.31
*P(Fault=***TRUE**|Day=**Thursday** « *Max Wind=***Greater than 12.45 ms^{-1}***)*	0.36
*P(Fault=***TRUE**|Day=**Friday** « *Max Wind=***Greater than 12.45 ms^{-1}***)*	0.31
*P(Fault=***TRUE**|Day=**Saturday** « *Max Wind=***Greater than 12.45 ms^{-1}***)*	0.36
*P(Fault=***TRUE**|Day=**Sunday** « *Max Wind=***Greater than 12.45 ms^{-1}***)*	0.36

Fault itself has an impact on two other variables, namely the likelihood of getting a Follow on Fault and the overall risk of a Fault. Clearly, both of these are manifests of the same condition. However, both have been included. Follow on Fault is a single event whereas Fault Risk captures a period of elevated risk [3]. The impact of Fault on both of these is consistent with raising risk due to intervention. The conditional probability table for Follow on Fault also contains an influence from humidity. This influence is an interesting one, from which we can speculate the physical nature of the relationship. There are some $6 \times 2 \times 2 = 24$ possible outcomes captured in this table, so only the salient features are given here. Follow on Fault is of course quite likely if Fault is TRUE, and likewise very unlikely if there is no precursor fault.

From Table 4.3, there is a clear pattern of increased probability of Follow on Fault for the middle categories of Min Humidity. At first glance, this seems strange. Intuitively we would expect the probability to increase almost monotonically with Min Humidity. However, a plausible and somewhat obvious explanation can be found, namely the work habits of the engineers.

Table 4.3   Extract from conditional probability for Follow on Faults occurring.

P(*Follow on Fault*=**TRUE**|*Fault*=**TRUE** « *MinHumidity*=**Less than 37**)	0.35
P(*Follow on Fault*=**TRUE**|*Fault*=**TRUE** « *MinHumidity*=**Between 37 and 53**)	0.59
P(*Follow on Fault*=**TRUE**|*Fault*=**TRUE** « *MinHumidity*=**Between 53 and 65**)	0.82
P(*Follow on Fault*=**TRUE**|*Fault*=**TRUE** « *MinHumidity*=**Between 65 and 70**)	0.72
P(*Follow on Fault*=**TRUE**|*Fault*=**TRUE** « *MinHumidity*=**Between 70 and 86**)	0.61
P(*Follow on Fault*=**TRUE**|*Fault*=**TRUE** « *MinHumidity*=**Greater than 86**)	0.50

It is highly unlikely that an engineer will open a pole top DP in heavy rain. In fact, the area manager has expressly forbidden engineers from doing so. However, engineers need to clear the faults as quickly as possible. Therefore, repairs are more likely to be attempted during a lull in weather conditions (but where ambient moisture is still high). This explains the maxima in this subset.

Finally we turn to the weather impact on the Fault Risk. This is a very large conditional probability table, with some $2 \times 5 \times 6 \times 7 = 420$ possible outcomes. Clearly, this contains a great deal of null information. In order to combat this we run a function which extracts the highlights and lowlights from the table to help better understand the relationships. From this, we can see there is very little to differentiate the probabilities for when Fault Risk is ordinary. Again, this is consistent with low risk.

However, although we see similar weather interaction as above when Fault = TRUE, there is a great deal of difference. This is because it is an important factor in fault propagation regardless of the weather. When *Fault* = FALSE, then the chances of a fault increase purely due to the weather itself. At this juncture, we point out that Max Wind has no influence on Fault Risk. Again, this is consistent because Fault

Risk is a temporally smeared variable and Max Wind a transient storm variable. Therefore, although it has an impact on point fault occurrences, we are unlikely to identify any relationship with Fault Risk. The reported highlights for these relationships are given in Table 4.4.

Table 4.4   Extract from conditional probability for Fault Risk.

*P(Risk=**HIGH**|Fault=**FALSE** « Rain=**2.5-5.5** « RainPast=**4** « Hum=**53-65**)*	0.44
*P(Risk=**HIGH**|Fault=**FALSE** « Rain=**2.5-5.5** « RainPast=**3** « Hum=**70-86**)*	0.52
*P(Risk=**HIGH**|Fault=**FALSE** « Rain=**2.5-5.5** « RainPast=**4** « Hum=**70-86**)*	0.54
*P(Risk=**HIGH**|Fault=**FALSE** « Rain=**5.5-13.5** « RainPast=**6** « Hum=**70-86**)*	0.44
*P(Risk=**HIGH**|Fault=**FALSE** « Rain=**2.5-5.5** « RainPast=**3** « Hum=**70-86**)*	0.52
*P(Risk=**HIGH**|Fault=**FALSE** « Rain=**2.5-5.5** « RainPast=**3** « Hum=**GT 86**)*	0.58

From Table 4.4 it is clear that Fault Risk is high for many conditions which provide for high moisture ingress. This can be achieved in a great many ways, and will of course depend on the plant type itself; we have a great number of possible outcomes from this table confirming this. The worst cases are given in Table 4.4.

4.6   Suggestions for Further Work and Summary

This chapter has illustrated a straightforward probabilistic modelling approach which provides a clear view of the fault propagation process within the copper access network. At this current level of application it enables a better understanding of the aetiology of faults alongside a scalable means of building multiple models over wide areas of the network. This in turn allows analysts to identify weak nodes and thus better employ scarce resource in upgrading and surveying the network, and pin-pointing hot spots for contingency provision in *force majeur*e. Furthermore, it is useful to compare models between areas, so enabling comparison of both underlying health and work practices. This would be helpful in better understanding differences in performance between areas, and addressing the issues relating to this. All of this has direct benefit within the current broadband delivery process, where customer satisfaction must be maintained in the face of possible loss of service.

As this is still a rough datamining tool, the authors recognise many potential areas of improvement and future work. Clearly, there is still much to be done in terms of clarifying the weather variables used in this study, and perhaps some further simplification could be made. However, despite this criticism we have presented a modelling approach which still outperforms any current offering. It is simple (in its realisation at least), and is certainly insightful and scalable. Although simple, it paints a very rich picture of how faults are propagated within nodes on the network, and even differentiates intelligently between weather aetiological factors.

It has been applied to underground DPs in a similarly successful way. However, we do clarify that not all DPs provide good fits. Although upon examination of these data series, it is hard to see how any approach would give insight. The respective DPs appear to be robust against the prevailing weather conditions and are also shown to be borderline in terms of intervention [3]. It is also possible to see how the addition of manpower and personnel data could further enhance the models, and provide further insight into how field engineering could be improved.

Current work relates to using these models as meta data, and the construction of approaches for data mining the models themselves to gain further insight into temporal and spatial trends and clustering patterns. Certainly, previous attempts using even simpler single measures [3] give spatial patterns and we hope to extend this work further.

Appendix

4A   A Data Structure for the Bayesian Network Model

The data input for the model took the equal frequency discrete values shown in Fig A1 — they are defined as:

- Fault specifies if the DP experienced a fault in the day in question;

- Follow on Fault states whether a fault occurred in the next ten days following this fault;

- Fault Risk temporally smears the fault reported by taking higher values in the ten and two days prior to a fault occurring;

- Day of Week is a nuisance variable stating on which day of the week the event occurred.

The remaining weather variables are straightforward daily summary measures:

- the Min Humidity acts as a baseline ambient moisture variable for the day;

- Rainfall is a daily total;

- Rainfall in Past is a measure of the number of days in the last six which had significant rainfall (more than 2.5 cm);

- Max Wind captures the highest gusts on that day.

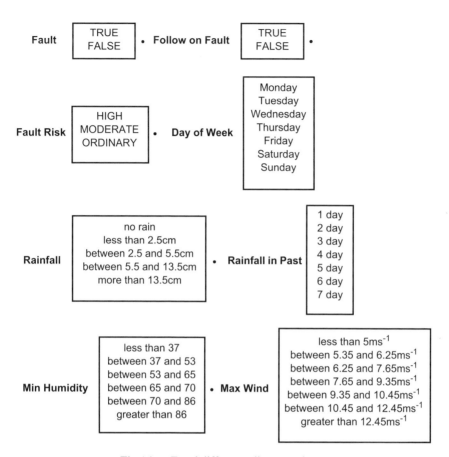

Fig A1 Equal difference discrete values.

References

1   Maxwell, K.: '*Residential Broadband*', Wiley (1999).

2   Thomas, M., Bell, P., Gould, C. and Mellis, J.: '*Fault rate analysis, modelling and estimation*', BT Technol J, **14**(2), pp 133-139 (April 1996).

3   Yearling, D. and Hand, D. J.: '*Classifying temporal patterns of fault behaviour within large telephony networks*', Intelligent Data Analysis Conference, Lisbon, Portugal (2001).

4   Pearl, J.: '*Causality: Models, Reasoning and Inference*', Cambridge University Press (2000).

5   Kenny, D.: '*Correlation and Causality*', Wiley (1979).

6   Gelman, A. et al.: '*Bayesian Data Analysis*', Chapman and Hall (1998).

7   Jensen, F.: '*An Introduction to Bayesian Networks*', UCL Press (1998).

8   Everitt, B. S. and Dunn, G.: '*Applied Multivariate Data Analysis*', Arnold (2001).

9   Lauritzen, S.: '*Graphical Models*', Clarendon Press (1996).

10   Whittaker, J.: '*Graphical Models in Applied Multivariate Statistics*', Wiley (1990).

11   Edwards, D.: '*Introduction to Graphical Modelling*', (2nd edition) Springer (2000).

12   Cheng, J., Bell, D. and Liu, W.: '*Learning Bayesian networks from data: an efficient approach based on information theory*', ACM Information and Knowledge Management (1997).

5

EMERGENT PROPERTIES OF THE BT SDH NETWORK

J Spencer, D Johnson, A Hastie and L Sacks

5.1 Introduction

When modelling synchronous digital hierarchy (SDH) networks it is usual to consider the algorithms responsible for the planning of these networks. These algorithms are well defined and consider a wide range of inputs describing the network scenario, including the demands of the network, the configuration of nodes, available physical layer capacity as well as restrictions placed on the design by various technological issues. During planning it is also common to incorporate deliberate structures such as rings to achieve additional attributes such as resilience — as we have already seen in Chapter 2. Hierarchies are also used in the network to make the management of capacity easier.

It would seem that the real-world result could therefore be easily deduced from knowledge of the demands and the planning algorithm used. In this chapter we will show that even with strict planning, emergent and unplanned topological traits appear in the real-world SDH network considered here. The existence of such traits within large-scale networks is not without precedence, however. It has been previously shown [1] that the Internet demonstrates similar traits at a number of levels, from router connectivity to the connectivity of sub-domains. In stark contrast, however, the Internet is not globally planned like an SDH network, does not have enforced structure and often uses dynamic routing techniques to adapt to network conditions. It may not therefore be surprising to find such emergent topological traits there.

This chapter begins by examining the overall telecommunications system and its constituent layers. It then briefly describes the macroscopic traits found in the Internet, an example of one of the layers, the Internet protocol (IP) layer, and then demonstrates the existence of these traits in a deployed SDH network. Additional traits to those found in the Internet are then illustrated, including some describing topology, bandwidth distribution and geographic connectivity. Possible explanations are then offered about to the source of these traits and some existing models that attempt to emulate these traits in the Internet are described, along with their applicability to SDH networks.

5.2   Multi-Layer Networks

To provide functional services to the end user, a telecommunications network consists of a number of layers, each contributing features required to provide the final service. SDH is the transport layer that gives a manageable interface to the physical capacity. It can be seen as part of a simplified multi-layer network, as shown in Fig 5.1. End users make demands that must be satisfied by the layers below, all of which impose limitations and restrictions, such as available resources, or technological limitations. For example, SDH circuits can be carried in wavelengths of which there are a finite number per fibre, and which can only

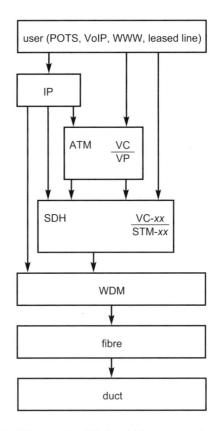

Fig 5.1   A simplified multi-layer network.

traverse available installed fibre. The fibre topology is similarly dictated by duct availability. The layers are therefore closely coupled and highly dependent on each other.

5.3   Large-Scale Networks

Of the many layers in such a multi-layer network, this chapter will consider only two, together with an example network of each.

5.3.1   The Internet

The largest example of an IP network is the Internet. It consists of a large number of interconnected IP networks that are not strictly planned and often considered near random in topology.

Internally the sub-networks use interior routing protocols such as open shortest path first protocol (OSPF) or routing information protocol (RIP) to route along the shortest number of hops or to minimise other routing metrics. Internetwork routing is governed by protocols such as the border gateway protocol (BGP) which implements a set of operator-defined policies to allow routing of traffic between autonomous system (AS) domains. The Internet has evolved and experienced huge growth but has no planned global structure or design and yet has been shown to exhibit a number of emergent topological traits. Faloutsos et al [1] examined a single instance of a router topology and a number of instances in time of the AS domain topology. They found that the following four power laws held true of all the topologies.

- Power-Law 1 (rank exponent)

  The outdegree (connections from a node) was found to be proportional to the rank of a node, to the power of a constant, the rank being the position of the node in a table sorted (numerically decreasing) by the outdegree of the node:

$$d_v \propto r_v^R \qquad\qquad ...... (5.1)$$

- Power-Law 2 (outdegree exponent)

  The frequency of an outdegree, f_d is proportional to the outdegree to the power of a constant, O:

$$f_d \propto d^O \qquad\qquad ...... (5.2)$$

- Power-Law (approximation) 3 (hop-plot exponent)

  The total number of pairs of nodes, within h hops of each other, is proportional to the number of hops to the power of a constant. This is more of an approximation since it only holds for values of h that are much less than the network diameter:

  $$P(h) \propto h^H, h \ll \delta \qquad \qquad \text{...... (5.3)}$$

- Power-Law 4 (eigenvalue exponent)

  The sorted eigenvalues (decreasing order), λ_i of the adjacency matrix (an N by N matrix (where N is the number of nodes) which is 1 when the two nodes are connected and 0 otherwise) are proportional to the index, i, into the list, to the power of a constant, ε. The power law was shown to hold for only the top 20 or so eigenvalues:

  $$\lambda_i \propto i^\varepsilon \qquad \qquad \text{...... (5.4)}$$

These very distinct traits are not planned and must exist as the result of individual design decisions by network operators or be the result of routing protocols. RIP is a distance vector routing protocol that attempts to optimise routes based on the number of hops, while OSPF is a link-state routing protocol that optimises for a specified metric (the default in Cisco routers for example is link bandwidth). They are both capable of reconfiguring the network in minutes, rather than hours or days.

5.3.2   SDH Networks

In contrast to the dynamic and relatively unplanned nature of the Internet, SDH transport networks [2] are meticulously planned on a global (macroscopic) scale. They provide end-to-end circuits with strict guarantees on the delivered levels of service (such as available capacity, delay or jitter). To provide these guarantees the network must be dimensioned and planned in advance with the demands on the network being known (or carefully forecast) beforehand. During the planning stage explicit structures, such as rings, are often enforced in the network to provide for the easier provisioning of resilience, as they are a simple way to provide alternate paths through the network.

Capacity is also often allocated to form a hierarchy, which allows for the easier management and better growth potential of the network. Any changes to the network are strictly planned and controlled and happen on a weekly to monthly scale.

5.3.2.1 BT's SDH Network Structure

These design principles can be seen in practice in the generalised diagram of the BT SDH network in Fig 5.2.

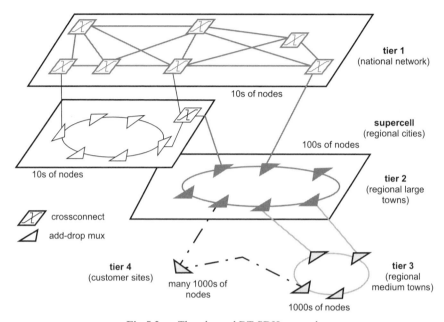

Fig 5.2 The planned BT SDH network.

The network consists of approximately four tiers, the lower tiers being of a ring nature, multi-homed into upper tiers, while the top tier is a collection of highly meshed nodes. However, the greatest influence on the location and connectivity of these elements is still by equipment and fibre availability, user distribution and collocation of adjacent layer equipment.

5.3.2.2 SDH Site Topology

To investigate the deployed topology and the possibility of emergent properties a single instance of the configuration of BT's narrowband SDH network was considered. The snapshot consisted of end-to-end circuits, the bandwidth allocated to these circuits and the geographic position of all the nodes traversed by the circuit. In this study we will be considering the connectivity of the sites, rather than individual rack equipment. In our network topology, a node is considered to be a site and two nodes are linked if there are one or more circuit hops passing between them. Since we are only considering the core network all customer nodes have been removed as well as the links to these nodes.

The network consisted of a few thousand nodes and had a links-to-nodes ratio of 1.77. A number of the plots in this study have been normalised to protect certain network information — the power-law fit exponents or correlation factors are not influenced but the *y*-intercepts are.

Figure 5.3 shows a plot of the degrees (the number of links connected to a node) of each of the nodes against their rank when sorted in decreasing numerical value of degree. The SDH network exhibits a relatively good conformance to the power-law fit with an R^2 value of 0.929 (R^2 is a standard statistical measure of deviation of a fit to data points with 1.0 being an exact fit). This is equivalent to Power-Law 1 in the Internet topologies. Degree rather than in-degree (number of incoming links) or out-degree (number of outgoing links) is being considered, as all circuits are symmetric.

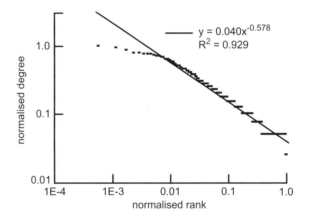

Fig 5.3   Power-Law 1 — the degree rank exponent.

In Fig 5.4 we have the frequency distribution of the degrees of the nodes (Power-Law 2). We can see that again there is good conformance ($R^2 = 0.932$) to the power-law fit. This suggests that the network has developed a tendency to have a very large number of nodes with low connectivity and a small number of nodes with a very high connectivity.

In Fig 5.5 we have a plot of the number of pairs that are separated by the given number of hops (Power-Law 3). This law is only an approximation as it only holds when a node pair separation is 7 hops or less.

Power-Law 4 in Fig 5.6 is the eigenvalue power-law which is a plot of the eigenvalues of the adjacency matrix sorted in decreasing order plotted against the index into that list. Eigenvalues are a function of the network size, the cliques (subgraphs) present in a topology, and their connectivity. The SDH data shows very good conformance to the power-law fit ($R^2 = 0.995$), even for the thirty highest eigenvalues.

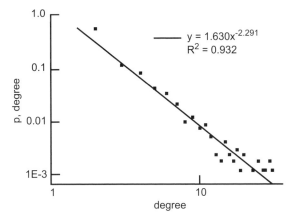

Fig 5.4   Power-Law 2 — the frequency distribution of node degree.

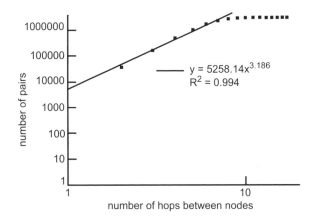

Fig 5.5   Power-Law 3 — the hop-plot approximate power-law
(only holds when number of hops between nodes is 7 or less).

We can see that, while created very differently, the Internet and this SDH network both exhibit the same emergent topological traits and follow the same power-laws, albeit with different exponent values. The SDH network does, however, follow further power-laws.

In an effort to find the effect of imposed structure it was found that the clustering co-efficients [3] of the nodes formed a power-law with their rank in a sorted list. The clustering co-efficient of a node is defined as the ratio of the total number of links between all of the nodes in the neighbourhood of a node (the node and all of its immediate neighbours) and the total number of possible links in that neighbourhood:

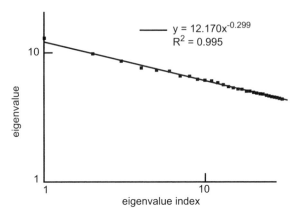

Fig 5.6   Power-Law 4 — the descending sorted list
of real eigenvalues of the adjacency matrix.

$$c_v = \frac{l_{neigh}}{0.5d_v(d_v + 1)} \qquad \text{......} (5.5)$$

where c_v is the clustering co-efficient of node v, l_{neigh} is the number of links between all nodes in the neighbourhood, and d_v is the number of nodes directly connected to the central node v.

The plot of co-efficient against rank can be seen in Fig 5.7. It has a number of interesting features, including the upturned tail on the left and the plateaus in the right half. The tail could be the result of the removal of customer nodes and links, which would cause an increase in the clustering co-efficient of edge nodes since they were previously hub nodes. The rightmost plateau, which is at a clustering co-efficient of 1.0, is probably the tier 1 nodes which are highly meshed. The much wider plateau of 0.66 is probably caused by homogeneous rings in the lower tiers as there so many of them.

The fifth power-law can therefore be defined as:

- Power-Law 5 (clustering co-efficient rank exponent)

  The clustering co-efficient, c_v, was found to be proportional to the rank of the co-efficient, to the power of a constant, C:

$$c_v \propto r^C_c \qquad \text{......} (5.6)$$

We have now seen how a globally planned network exhibits a range of topological traits. To further understand the sources of these traits and the extent to which they are present we will consider the geographic distribution of the SDH sites.

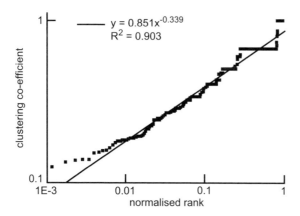

Fig 5.7   Power-Law 5 — the rank plot of the clustering co-efficients of all the nodes.

5.3.2.3   Network Geography

One of the aims of network planning is cost minimisation and, as a result, network access points are often placed as near to the centre of clusters of customers as possible. This minimises costs associated with customer access systems and therefore causes the geographic site distribution to follow population distribution. It has previously been shown that the customers, and therefore SDH nodes, follow a fractal distribution [4], but we must examine the effect that the distance between SDH sites may have on their connectivity. In Fig 5.8 we find the distance between every node pair plotted against the probability of a link existing between nodes over that range. The distances between nodes are grouped in bins of 500 distance units.

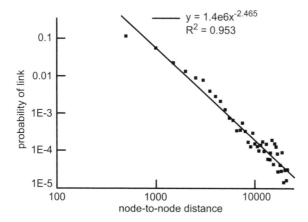

Fig 5.8   The probability that a pair of nodes are connected versus the geographic distance between the two nodes (distances have been put in 500 distance unit bins).

We can see that there is a sharp drop in linking probability as the distance between nodes increases. Since there is a clear relationship between link probability and distance, the connectivity must be a function of node distribution and therefore physical layer cost functions.

Now we will consider how prevalent the traits are throughout the network geography and the network hierarchy. As we move from town, to city, and then to national scale, we follow the tiers up the hierarchy of Fig 5.2. To examine the existence of these traits throughout the tiers we can divide the country by a grid of fixed-size squares. The width of the square will be denoted by M, which is in the distance units used before. As M increases, more and more nodes are grouped within these squares. The connectivity of the squares then follows the connectivity of the nodes enclosed within the squares. As M increases, each square will encapsulate towns, then cities, and counties, and finally the whole country will be contained within a single square.

In Fig 5.9 we see the effect of increasing M on the exponent of the degree distribution (Power-Law 2). The right axis shows the R^2 value of the power-law fit. It is not until M reaches 4000 distance units that the conformance drops below 0.90. At this scale we are examining entire towns, but large cities could still span multiple grid squares.

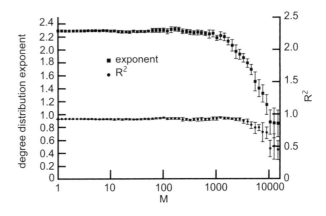

Fig 5.9   The Power-Law 2 exponent for increasing grid size, M, as well as the conformance to the power-law fit, R^2, on the right axis.

So even the connectivity of entire towns, containing many SDH sites follows Power-Law 2. The extent of this can be further seen in Fig 5.10 where R^2 is plotted against M, but with the fraction of total nodes in the grid on the right axis. The conformance does not drop below 0.90 until the topology contains 10% of the original number of nodes.

The grouping of nodes into squares at different scales (values of M) is equivalent to the box-counting method [5] for finding the fractal dimension of the geographic node distribution. The clear slope between M = 1000 and M = 10 000 demonstrates that site distribution is scale invariant and the gradient of this slope, and therefore the fractal dimension, is 1.46.

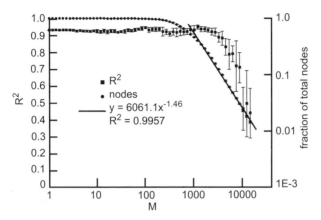

Fig 5.10   The conformance to Power-Law 2, for increasing M, with the fraction of total number of nodes formed by the grouping of the grid, on the right axis.

5.3.2.4   *Bandwidth Distribution*

Another aspect of the network to be influenced by network geography and topology is bandwidth distribution. User distribution, and therefore demand end-points, together with routing algorithms and the available physical capacity, all dictate the distribution of allocated bandwidth within the network. In Fig 5.11 we can see this distribution, where total allocated capacity between nodes, rather than available physical capacity is plotted.

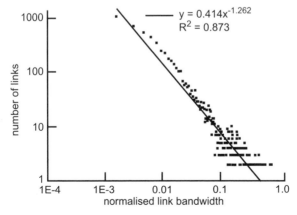

Fig 5.11   Allocated bandwidth distribution between sites (normalised bandwidths have been put in fixed-size bins).

This distribution follows a power-law with a weaker conformance of 0.873 and has a slight tendency to arch on the left half of the plot.

We have now seen how a globally planned network, designed by well-specified algorithms and under the influence of undefined forces, has formed a number of macroscopic emergent traits. The traits do not appear explicitly in network design algorithms and therefore must be the response of the algorithm to the input data set.

5.4   Where do these Emergent Properties Originate?

The emergent traits of the SDH layer are unplanned and unexpected. The source of the traits must therefore lie in the formation and evolution of the network and unspecified local effects on the growth. SDH networks do not use dynamic routing and when changes do happen they occur on weekly or monthly scales and are often incremental rather than entire reconfigurations. The major influences on these changes are fourfold.

- Technology

   There are very large restrictions on the planner, from equipment limitations, to capacity limits, capacity granularity, physical connectivity and legacy network demands. The BT SDH network also has to cater for the large amount of legacy traffic that was originally plesiochronous digital hierarchy (PDH) based. The original planning of this traffic was decided by PDH planning techniques that had even more technological restrictions, such as those imposed by the 'multiplexer mountain'. In contrast to this, IP networks are much more dynamic and far less inhibited, with links capable of crossing the entire country through a single SDH or ATM circuit and there being no capacity granularity issues. IP dynamic routing protocols, which can reconfigure the network in seconds or minutes, often have metrics based on either link capacity (default OSPF metric behaviour in Cisco routers) or hop count (RIP) to destination. SDH planners consider a combination of the two. These technologies, routing protocols or network architectures, do not, however, impose explicit power-law traits.

- Network architecture

   When adding new links to an existing SDH network the planned structure must be followed. It can be shown, however, that simple models of an SDH network such as a three-tier random collection of rings of nodes does not exhibit such laws, neither does a three-tier collection of networks which are internally randomly connected (a simple model of the Internet). Aside from the basic connectivity elements (rings, meshes) the two network types differ further — in organisation. The Internet consists of stub networks that contain end users, and transit networks that carry the traffic between the stub networks [6]. In SDH networks on the other hand 97% of the sites are connected directly to an end customer; there is little distinction between edge nodes and core nodes. It is conceivable therefore that the topology of the SDH network is much more closely coupled to the demand of the layers above than an IP network would be.

- Adjacent layer cost functions

  The design of each layer must be accommodated by the layer below and as such the design will be influenced by cost functions imposed by that layer. Due to its proximity to the physical layer SDH is much more tightly coupled to geography. IP on the other hand makes demands of the transport layer and is less inhibited by the physical layer — a single hop in an IP network could be carried by a single SDH circuit across the length of the entire country. IP does not necessarily need to be carried over SDH either, and could use a range of layer 2 technologies, such as frame relay, FDDI or Ethernet, all of which have their own limits.

  It can be shown though that cost-function-based connectivity models for geographic distribution do not necessarily produce power-law compliant topologies. Based on uniformly random node distribution, or even heavy-tailed node distributions and a connection cost function such as that proposed by Waxman [7], the traits do not exist.

- Demand growth characteristics

  The specifics of the demands being added could result in these traits. Different types of customers will exhibit different types of growth. With the growth of Internet usage at home, residential customers may require more capacity to reach Internet service provider (ISP) points of presence (PoPs), or, with the growth of virtual private network (VPN) services, there will be increased demand between business sites. If the demand problem is simplified to assume that the new nodes will have an affinity to the larger of the existing nodes, then we could model such growth using a prototype by Albert and Barabási [8]. Their models concentrated on preferential connectivity according to existing connectivity; new nodes would connect to existing nodes favouring the more connected nodes. The models do in fact produce topologies that follow power-laws. The exponents of the fits are, however, different to the exponents measured for the Internet and BT SDH network, but, more importantly, they are too simple as they totally disregard cost functions such as those associated with geographic distribution and technological restrictions. Additionally, for SDH networks, they do not consider the explicit network design.

Of these possible influences only the growth model by Albert and Barabási could produce the power-law traits. The effect of the other influences is undeniable however — the network architecture does exist, as do cost functions (Fig 5.8) and so do technological limitations. For further models producing power-law topologies, see Mitzenmacher [9], and on the effect of structure on topologies, see Tangmunarunkit et al [10].

5.5   Self-Organising Criticality and Multi-Layer Feedback

When discovering the emergent traits of the Internet, Faloutsos et al suggested [1] that the source could be a self-organising system that would reconfigure the topology to cater for increased network demand. This could imply that core nodes with their higher capacity links will also have higher connectivity. To investigate this, Fig 5.12 has a plot (for the SDH network) of the bandwidth of a link against the degree of the less connected of the two end-points of the link. There is a very concentrated cluster in the lower left quadrant of the graph and a lack of high bandwidth with low degree links; the correlation co-efficient of the points is 0.611.

This would suggest that, if there is a relationship between node size (capacity) and its connectivity, it is not a simple one. This being the SDH network, the lack of an obvious relationship could be caused by imposed structure or the lack of a clear distinction between edge and core nodes. The lack of a clear relationship does not, however, discount the theory; it can be shown that the addition of core links to sustain increased demand from a growing network also creates power-law adherent networks [11].

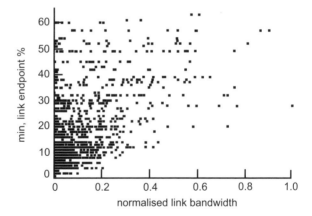

Fig 5.12   The lower of the two degrees of the link end-points versus the allocated
bandwidth of the link.

The problem again with this growth model is the lack of physical layer cost functions and network structure. In Spencer and Sacks [11], the use of physical layer cost functions immediately made the network deviate from the power-law traits.

The source of the traits is the subject of on-going research and the contributions of all of the above sources must be considered. The universality of the traits, being in different layers and at various levels (i.e. it exists in both Internet router and AS domain topologies) of each network suggests the source may lie in a more inherent process.

It has been suggested that as part of the self-organising system the need to minimise resource usage could be responsible [12]. That would certainly be present in both SDH networks, as part of the planning algorithm, and the Internet, as part of metric-based routing protocols. If the dynamics of layer growth and its effects on adjacent layers are to be properly understood, then better models for these traits must be devised.

5.6   Discussion and Summary

This chapter has presented a number of unplanned traits emerging from a strictly planned and highly structured SDH network. The traits are present at the site topology level and continue up to the connectivity of towns and cities. While this was an examination of only a single instance of an SDH network, the mere possibility has very wide reaching implications.

In traffic forecasting and scalability analysis, knowledge of the macroscopic design can be used to find bottle-necks and under-utilisation. These traits also give a better picture of the demand an SDH network can place on the physical layer, and WDM planning can now benefit from more realistic demand models, as would simulation of the SDH network itself [13]. If the sources of these traits were properly captured then it would be possible to predict the effect of physical layer changes on the layers above and develop more robust and scalable networks.

The actual traits, and not just modelling based on them, have wide ranging implications too. As there are no size-dependent characteristics in the plots, the network is considered as not having obvious limiting factors to its scalability. The precise source of the traits is elusive and is conceivably part of the large-scale problem of resource minimisation within multi-layer networks and the feedback between the constituent layers. Layer limitations, design principles, legacy network contributions and population interaction, all play a part in this self-organising feedback system and their contributions are hard to evaluate.

References

1   Faloutsos, M., Faloutsos, P. and Faloutsos, C.: '*On power-law relationships of the Internet topology*', in Proceedings of SIGCOMM: IEEE Conference on Communications (1999).

2   Brown, R. S., Rowland, D., Vinall, A. and O'Neill, A.: '*Broadband transport —the synchronous digital hierarchy*', BT Technol J, **16**(1), pp 148-158 (January 1998).

3   Watts, D. J. and Strogatz, S. H.: '*Collective dynamics of small-world networks*', Nature, **393**, pp 440-442 (June 1998).

4   Appleby, S.: *'Fractal telecommunication networks'*, BT Technol J, **12**(2), pp 19-29 (April 1994).

5   Peitgen, H. O., Jürgens, H. and Saupe, D.: *'Length, area and dimension: Measuring complexity and scaling properties'*, Chapter 4 in: *'Chaos and Fractals: New Frontiers of Science'*, New York, Springer-Verlag (1992).

6   Calvert, K., Doar, M. and Zegura, E.: *'Modelling Internet topology'*, IEEE Communications Magazine (June 1997).

7   Waxman, B. M.: *'Routing of multipoint connections'*, IEEE Journal on Selected Areas in Communications, **6**(9), pp 1617-1622 (1988).

8   Albert, R. and Barabási, A. L.: *'Topology of evolving networks: local events and universality'*, Phys Rev Lett, **85**(24), pp 5234-5237.

9   Mitzenmacher, M.: *'A brief history of generative models for power law and lognormal distributions'*, 39th Annual Allerton Conference on Communication, Control, and Computing (2001).

10 Tangmunarunkit, H. et al: *'Network topology generators: degree-based vs structural'*, SIGCOMM, IEEE Conference on Communication (2001).

11 Spencer, J. and Sacks, L.: *'Modelling IP network topologies by emulating network development processes'*, IEEE International Conference on Software, Telecommunications and Computer Networks (2002).

12 Fabrikant, A. et al: *'Heuristically optimized trade-offs: a new paradigm for power laws in the Internet'*, International Colloquium on Automata, Languages, and Programming (ICALP2002)(July 2002).

13 Anagnostakis, K., Greenwald, M. and Ryger, R.: *'On the sensitivity of network simulation to topology'*, Proceedings of the 10th IEEE/ACM Symposium on Modelling, Analysis, and Simulation of Computer and Telecommunications Systems (MASCOTS) (2002).

6

EMC EMISSIONS CERTIFICATION FOR LARGE SYSTEMS — A RISK-MANAGEMENT APPROACH

D J Carpenter

6.1   Introduction

6.1.1   What is EMC?

Electromagnetic compatibility (EMC) is the field of engineering concerned with the ability of electrical and electronic devices to co-exist with users of the radio spectrum.

EMC addresses two different issues that arise from the same phenomena within electrical and electronic devices — their unintentional antenna action. An antenna is a device specifically designed to do two things — either convert an input voltage signal to an output electromagnetic wave (when 'transmitting'), or alternatively to convert an input electromagnetic wave to an output voltage signal (when 'receiving'). An intentional antenna does this by design (its form being designed to maximise the efficiency with which it performs) through the exploitation of basic physics. Electrical and electronic equipment can also do this, as an unintended by-product of their operation.

To display unintended antenna action, an electrical or electronic device needs to have two components — an unintentional voltage source and an unintentional antenna.

An example that many readers will have personal experience of is the effect that a domestic, mains-powered hair dryer can have on Medium Wave (MW) broadcast reception. The electric motor within the hair dryer can, without due consideration during design, induce signalling onto the attached mains cable that in turn acts as an unintentional antenna. In this instance, the frequencies of the noise generated by the motor and the length of the main cable can conspire, totally unintentionally, to generate an efficient but again unintentional antenna. This has the effect that the EM field radiated within the vicinity of the cable can exceed the local levels of broadcast services, resulting in the degradation of reception quality. Depending upon the level generated by the hair dryer and the local level of the broadcast signal, the degradation can vary between the 'minor' (in which the received audio has an audible, perhaps annoying background noise level) through to the severe (in which the audio is completely unintelligible). Similar problems were once common with hand-held, mains-powered power tools such as electric drills.

Electronic devices can also display this behaviour. The tremendous increase in processor speeds and data transmission rates witnessed during the previous decade now means that digital circuits frequently contain many signals at radio frequency (RF). This is because the generation of stable and high frequency clock and data signals in the time-domain requires a series of harmonically related signal frequencies. To generate a 100 MHz square wave clock signal in the time-domain, for instance, requires signals at frequencies of 100 MHz, 200 MHz, 300 MHz, etc; that is all integer harmonics of the basic clock duty cycle. The signalling frequencies present on a printed circuit board (PCB) due to clock and data signals can conspire with either track lengths or attached data cables, again totally unintentionally, to generate an efficient but unintentional antenna.

The EM signals generated unintentionally by device operation are referred to within the EMC community as the device's radiated emissions. This terminology is intended to avoid confusion with the terms 'broadcasts' (intentionally generated EM signals, specifically designed to provide service over a defined area) and 'radiation' (this term being associated with atomic nuclear phenomena, referring to the α and β particles and high energy γ rays produced during the fission of atomic nuclei).

Both electrical and electronic devices can also demonstrate unintentional antenna action that leads them to act as unintentional receivers. In this instance, some facet of the device (such as an attached data or power cable or internal PCB track) acts unintentionally as an efficient receiver of EM signals within the local EM environment. This is a particular problem for digital electronics, since any unintentional receiver can introduce 'phantom' signalling into the device. This signalling is completely unrelated to any intentional signalling that is generated as part of the device's intended function, but can reach the correct voltage level to cause digital logic to respond. In this case, the device is likely to malfunction in some manner. Indeed, it is often wondered within the EMC community how many intermittent device faults that are assumed to be due to software bugs are in fact due to this unintentional reception behaviour.

The ability of a device to operate without malfunction in the presence of its local EM environment is referred to within the EMC community as the device's immunity. Sometimes the term susceptibility is also used. The two terms are reciprocal in meaning, since a device that is highly susceptible to malfunction due to

EM fields within its local environment clearly displays low levels of immunity, and vice versa.

6.1.2   Why Should BT be Interested in EMC?

6.1.2.1   Necessity

As a network operator, BT will deploy electrical and electronic devices in many different locations, each of which will have a different EM environment. The nature of the EM environment, and hence the facet of EMC (emissions or immunity) that is of interest, depends upon a device's location.

If the device is located very near a broadcast antenna, the device's local EM environment is dominated by the high EM field levels generated by the broadcast antenna. In this position, the main concern is one of immunity: whether the device can operate and deliver its intended function without interference due to the presence of the local high-EM field level. The emissions are of far less concern, because the field strength of the broadcast antenna is far greater than any generated by the unintentional antenna action of a device, hence the likelihood of interference to neighbouring receivers of the broadcast service is minimal.

If, however, the device is located far from the broadcast antenna, in a region where the broadcast signal will be quite weak, the main concern is one of emissions: whether the device can operate and deliver its intended function without causing interference to neighbouring receivers of the broadcast signal (that has decayed due to propagation to a relatively low level). The immunity of the device is of far less concern, because the broadcast field strength is so low that any 'phantom' signals generated through unintentional antenna action are very unlikely to be of a high enough level to cause erroneous switching and hence malfunction.

In reality, a deployed electrical or electronic device will be located in an EM environment that contains many such signals from a variety of different broadcast antennas. Hence BT will experience both immunity and emissions issues within its extensive network that need to be addressed in order to guarantee reliable network operation and related quality of service (QoS) levels. Indeed, BT Exact's EMC consultancy has accumulated enormous experience of resolving such issues, resulting from the vast variety of devices that are installed within a network that spans so many very different EM environments.

6.1.2.2   Regulation

Clearly, as a receiver is located further from a broadcast antenna, the signal level received decreases. There will plainly come a point at which the level of local, unintentionally generated emissions will begin to cause interference to reception quality. Obviously, if local, unintentionally generated, emissions are kept to an absolute minimum, the distance from the antenna at which interference occurs is maximised, and vice versa. However, controlling emissions generally introduces cost to electrical and electronic devices — the tighter the control, the greater the cost.

Historically, national governments throughout the developed world have introduced a series of statutory requirements on device emissions in order to provide a set of minimum assumptions that radio users could use in service planning.

At present, most of these requirements derive from the International Electrotechnical Commission's (IEC) Comité International Spécial Des Perturbations Radioélectriques (CISPR — generally pronounced 'sisper') — or International Special Committee On Radio Interference.

Within Europe, the regulatory framework is defined by the European EMC Directive (specifically 89/336/EEC[1] — Electromagnetic Compatibility (EMC) and 91/31/EEC[2]).

6.1.3   An Overview of the EU EMC Directive

The Single European Market (SEM) was created on 01-01-92. The market is intended to deliver freedom of movement of people, goods and capital across the national boundaries of its member states. The European Commission (EC) has since been working towards the implementation of a set of harmonised technical requirements that replace previous national requirements and thereby facilitate the desired free movement of goods.

The EMC Directive is part of this harmonisation process. It is also an example of a 'New Approach' Directive, in that it defines first a series of essential protection requirements and then a regulatory framework through which compliance is to be enforced.

Essentially, the Directive recognises four basic classes of device.

- Components

  Defined as being a set of passive devices that have no intrinsic functionality themselves (examples include resistors, capacitors), components do not, as such, have to meet the essential protection requirements of the Directive.

- Apparatus

  Defined as being a set of components brought together for the purpose of delivering some required functionality, apparatus can be placed on the market for sale as individual items. Apparatus must meet the essential protection

---

[1] 89/336/EEC had an original implementation date of 01-01-92.
[2] 91/31/EEC modified 89/336/EEC to delay implementation until 01-01-96.

requirements of the Directive. Responsibility for declaring compliance is on the party that places the apparatus on the market (generally the manufacturer for apparatus manufactured within the SEM, the importer for apparatus manufactured outside the SEM).

- Systems

  Defined as being items of apparatus that have been brought together and interconnected for the purpose of delivering some required functionality, systems can either be placed on the market for sale as individual items (as apparatus) or brought into service. Systems must meet the essential protection requirements of the Directive. Responsibility for declaring compliance is on the systems integrator (SI), the party that either created the system and placed it on the market (as apparatus) or brought the system into service.

- Installations

  Defined as being a set of collocated but not interconnected systems, installations do not, as such, have to meet the essential protection requirements of the Directive.

The Directive offers two basic routes to compliance.

- Standards

  The European standardisation body CEN/CENELEC is responsible for generating harmonised standards that embody the essential protection requirements of the EMC Directive. CENELEC publishes a list of issued harmonised standards, or Euro Normes (ENs), within the Official Journal of the European Commission (OJEC), that embody the protection requirements for different products. Where a relevant EN is available for a given product, the product may be tested in accordance with the EN and, if passed, is declared to meet the essential protection requirements. As such, all standards published by CENELEC in the OJEC have a presumption of conformity with the essential protection requirements of the Directive.

- Technical Construction File (TCF)

  When an EN is not available, it is still possible to demonstrate that a product complies with the Directive through the production of a TCF. This is essentially a document (a file) that contains all evidence and associated arguments that leads to the conclusion that the essential protection requirements are met. The TCF is essentially an exercise in due diligence, being documentary evidence that a party has given due regard to its legal obligations. Once produced, a TCF has to be submitted for approval to a Competent Body (CB), a recognised expert in EMC who has sufficient technical authority to comment on the evidence and argument presented.

Within the UK, the EMC Directive has been adopted on a complaints-driven basis. In the event of a complaint of interference to radio reception being reported, the UK regulator (being the Radiocommunications Agency (RA)) is obliged to investigate. In the event of a **product** being found to be causing interference and failing to meet the essential protection requirements of the EMC Directive, the regulator has the legal powers to withdraw the product from the market. In the event of a **system** being found to be causing interference and failing to meet the essential protection requirements of the EMC Directive, the regulator has the legal powers to enforce both:

- the shut down of the system for up to 6 months;

- the remedial engineering of the system to remove the source of interference.

6.1.4   The Impact of the EMC Directive on BT

Clearly, it is essential that BT ensures that any electrical or electronic device procured meets the mandatory market-entry EMC requirements. However, the inclusion of systems within the framework of the Directive has an additional and quite profound impact. This is because BT's business is essentially based upon deriving revenue from supplying voice and data services over its network. This revenue stream depends on the successful operation of a diverse range of physically large and highly complex switching, transmission, network management and support systems. In the provisioning of these systems, BT typically acts as an SI, integrating a wide variety of commercial off-the-shelf (COTS) hardware and software products from multiple vendors.

For systems containing a large volume of modern telecommunications and information technology equipment (ITE) items of apparatus, the radiated emissions present a particular problem. This arises from the existence of common emission frequencies among the items of apparatus. This is a particular problem when many similar apparatus items are integrated together. It is also a problem when different apparatus items are integrated together, because common signalling (and hence frequencies) — essentially those associated with synchronisation clocks and data exchange — will be processed throughout.

The fundamental problem is this: the Directive demands that both the individual apparatus and the composite system meet the same essential protection requirement of not causing interference to neighbouring radio receivers. While the apparatus items can be individually compliant to this requirement, the existence of many compliant emissions at common emission frequencies has the potential to produce a much higher, cumulative emission level for the system as a whole.

The fact that emissions naturally propagate beyond their point of production means that any emissions produced by a system will readily propagate beyond the system's premises — typically an exchange building. The fact that the majority of telecommunications systems are located within populated areas increases the probability of a radio receiver being located nearby.

Hence the combination of:

- the potential for 'high' system emission levels to occur at common emission frequencies;

- the potential for a radio receiver to be located near to the system;

- the number of systems deployed within BT's network,

indicates that there is an intuitively high probability that interference may occur.

Given the impact that a substantiated interference case would have upon the business (i.e. the loss of revenue due to the enforced shut-down of the system and the additional damage this would cause to the BT brand), it is clear that some strategy is required to manage this problem.

This chapter describes an EMC compliance strategy developed by BT Exact, reviews the options that were available to BT and identifies the commercial and technical difficulties associated with each. The chapter then goes on to present an overview of the strategy that has been developed in response to these difficulties.

6.2   Compliance Strategies — Options

6.2.1   Overview

The absence of appropriate harmonised standards containing technical requirements (i.e. test methods and limits) specific to large systems along with the difficulty of re-engineering should non-compliance occur has led BT to reject the standards route to compliance. Instead, BT has chosen to take the TCF route.

The method employed is required to meet a number of requirements specific to a network operator. Within the telecommunications domain, it is typical for a given system to be deployed within the network in a number of different configurations. The configurations vary in terms of their complexity, physical size and numbers deployed. At the one extreme there is the configuration that handles a small number of customers (e.g. the local concentrators that service small, rural communities) that will typically be deployed in large numbers. At the other extreme there is the configuration that handles very large numbers of customers (e.g. the major urban switching centres) that will typically be deployed in small numbers. Essentially, the dendritic nature of telecommunications networks gives rise to a pyramidal relationship between complexity, physical size and deployment numbers — as complexity and physical size increases, the deployment number decreases. The TCF will need to provide evidence that each deployment configuration meets the essential protection requirements of the EMC Directive.

Given the numbers in which each configuration is deployed within the network and the need for the TCF to cover all deployed configurations, a statistical approach is essential, as it is not economically viable to assess every deployment individually. By treating the deployment of each configuration as a statistical population, it is possible to employ sampling methods to minimise the number of deployments that need to be assessed while maximising the statistical confidence in the derived compliance status of the population.

The use of statistical sampling techniques within the EMC domain is well established by CISPR 22 [1]. This is equivalent to EN 55022, the harmonised standard carrying presumption of conformity under the EMC Directive for radiated (and conducted) emissions from an item of ITE apparatus. The document contains a method by which manufacturers may acquire statistical confidence in the compliance of their entire production run from the testing of a limited sample of devices from that run. The method requires that a small number of items of apparatus selected at random from a production run be tested. For those frequencies at which the highest emission levels are recorded, compliance of the production run is achieved through satisfaction of the following condition:

$$X + Sk \leq L$$

where:

L is the limit level,

X is the arithmetic mean of the measured emission levels at a specific frequency,

S is the standard deviation,

k is a defined factor corresponding to the condition that 80% of the population (of manufactured apparatus) can be expected to fall below L with 80% statistical confidence.

The method is therefore referred to as the 80-80 rule.

Hence the 80-80 rule provides an ideal method for BT to certify the radiated emission performance of its population of deployed systems. To do so, it is necessary to obtain radiated emission data on a statistical sample of deployments.

There were initially three options identified to acquire this information:

- *in situ* survey of deployed systems;

- site-based testing of example systems;

- mathematical modelling.

Each option was thoroughly analysed to determine its commercial and technical impact. This analysis is summarised within the following subsections.

6.2.2   *In Situ* System Emission Level Measurements

This option requires that a statistical sample of each deployed configuration be selected for radiated emission measurement via on-site survey. Such an approach presents a number of logistical and technical challenges.

The logistical challenges arise from the need to survey each selected deployment regularly over a period of time, to determine each deployment's individual emission characteristics. This need arises from the fact that all measurements on a live system yield emission levels associated with the system loading (i.e. the percentage of the capacity that is active) during measurement — as loading varies considerably over time, it can be expected that emission levels will vary similarly. Hence to capture emission levels associated with different system loads, it is necessary to make multiple measurements over an extended period of time, at different times of day and at different times of year.

This means that the measurement sample for each deployment can take a considerable time to accumulate. If a total sample of 50 deployments (covering all deployed configurations) are to be surveyed regularly and it is possible to perform two *in situ* surveys per working week, then the complete sample is surveyed once every 25 working weeks. This means that each deployment would be surveyed twice a year.

This is a classic cost-benefit trade-off. Increasing the number of surveys that can be done increases the resource demanded and hence cost, but decreases the time taken; whereas reducing the number of surveys that can be done decreases the resource demanded and hence cost, but increases the time taken. The resources required for this approach are not trivial, consisting of dedicated survey vehicle(s) equipped with the necessary measurement equipment (antennas, test receivers, etc) and trained personnel.

This approach also requires that the deployment remains stable over time. Throughout the latter half of the 1990s, following the implementation of the Directive, this was almost impossible to guarantee: at that time, the explosion in demand for telecommunications services (particularly Internet dial-up capacity) was driving an aggressive hardware deployment programme. This meant that any deployment selected for repeat survey was highly likely to have changed in some way between successive surveys. Hence the size of the data set applicable to each deployment was regularly being reset to one (i.e. that obtained during the most recent survey, following the most recent change to the deployment).

The technical challenges arise from the fact that the RF emission measurements are made in uncontrolled conditions. For instance, the measurement location (typically a telecommunications building) possesses unknown ground conditions that will vary considerably between deployments and systematically influence the measurements taken in a manner for which it is difficult to make correction. The measurement location will also typically contain a large number of ambient RF emissions due to sources other than the target system. These emissions come from outside and inside the measurement location and include local users of the RF spectrum and the other systems and items of apparatus that are deployed within the same building. The range of such apparatus items is enormous in terms of their functionality, physical size and age. Older apparatus is a particular issue, since this is likely to have been engineered to different emissions requirements than modern apparatus. The measurement location is also typically 'cluttered', containing many conducting structures near to the target system that, by acting as sources of reflection, could systematically influence the measurements taken in a manner for which it is difficult to make correction.

Hence it is generally necessary to perform measurements at many points around the perimeter of the target system. At each point an extensive investigation is typically required to identify the source of all identified emissions to confirm that it is due to the target system.

The analysis therefore concluded that while acquisition of the required emission sample through *in situ* survey would be possible, the results would be subject to some uncertainty and the sample would be likely to take several years to accumulate.

6.2.3   Site-Based System Emission Level Testing

This option requires that the radiated emissions from a statistical sample of each deployed configuration be measured at a test site. This has strong technical advantages over the *in situ* approach, since it allows a system to be constructed and exercised in a controlled manner, allowing the variation between emission level and system loading to be systematically investigated. It also ensures that all systems are tested on the same, standardised measurement facility, allowing the ready comparison between results. However, this approach presents a number of commercial challenges, since the testing of each member of the sample has a set of six significant cost components.

The first component arises from the need to build and maintain a facility large enough to accommodate the systems to be tested. The second arises from the need to transport both apparatus and exercising equipment to and from the facility. For testing to be practical, the delivery schedules for all items of apparatus and exercising equipment has to be synchronised — something that is often practically difficult when working with multiple vendors.

The third cost component arises from the need to install and commission the apparatus and exercising equipment at the test facility, then to remove once testing is completed. Given the need to consider all deployment configurations, this activity will occur throughout execution of the test plan. Given the complexity of the systems of interest, configuration is a non-trivial task that can easily consume weeks of effort (particularly for the very large systems).

The fourth cost component arises from performing the actual testing. Each deployment configuration will need to be tested, typically involving measurement at several points around the system perimeter (since the system's physical size is likely to rule out the use of a turntable). The need to consider many configurations and typically many positions around each combine to generate a test plan that can take weeks to systematically execute.

The fifth cost component arises from the need to have expert personnel available throughout execution of the test plan, to correctly configure the system and confirm correct operation during the test.

The sixth cost component arises from any losses associated with execution of the test plan. Essentially, tying up a large number of items of apparatus and exercising equipment during test has a cost — for vendors, any apparatus supplied is not in the supply chain and generating revenue from sales, while, for the SI, any apparatus purchased is not generating revenue.

The analysis therefore concluded that the acquisition of the required emission sample through site-based measurement would be prohibitively costly and hence not cost effective.

6.2.4   Mathematical Modelling

This option is extremely attractive, as it has the potential to deliver the required emission sample in a way that could avoid the various challenges presented by the other two options.

However, this approach presents its own set of technical challenges, as it essentially requires the construction of a set of accurate and detailed three-dimensional computer models for the various deployed configurations and their analysis using one of the available methods (such as the method of moments (MoM), the unified theory of diffraction (UTD), the finite difference time domain (FDTD) method, etc).

The need to generate accurate computer models presents a number of challenges. The process generally requires the amassing of considerable data, typically from disparate and often non-electronic sources. It also requires knowledge of the various modelling guidelines that arise from the detailed numerical method that will be employed by the code that will analyse the structure. Also, if significant variation exists among the detailed physical layout of a given configuration, it will be necessary to generate a number of models for that configuration, each of which will need to be analysed. Model construction is therefore a time-consuming and expensive undertaking.

Similarly, analysis of the models presents a number of technical challenges. These essentially derive from the fact that physically large systems are also electrically large (i.e. their physical dimensions are typically much larger than the wavelength of interest). Hence techniques (such as MoM and FDTD) that rely on a model containing electrically small elements generate very large computational problems that require considerable resource (primarily storage capacity and run-time) to process. Despite the ongoing advancement in processing speed and addressable memory size, many systems will generate computer models that are simply too large to process using a high-performance PC within an acceptable timeframe. The need to process a number of models (for the different deployed configurations) a number of times (to investigate the effect of system loading) simply multiplies the total run-time of the problem set.

The analysis concluded that the conventional mathematical modelling methods (such as MOM, UTD and FDTD) did not offer a practical strategy to acquire the required emission sample.

6.3   BT's Strategy

It was concluded that none of the three options analysed would provide a cost-effective method of acquiring the required emission sample.

To address this situation, BT developed a system-level emissions compliance strategy based upon risk management. This exploits an innovative, high-level mathematical modelling approach that is supported by on-site measurement. The approach developed naturally from consideration of a conventional high-level approach to the prediction of the system emission level.

6.3.1   System Emission Level Prediction — a Conventional Approach

When a system contains a number of apparatus items that each individually emit at some known level (when measured at some standard distance) at a common frequency, mathematical tools exist to predict the combined, system-level radiated emissions.

Imagine we have a number, N, of independently radiated emissions at some common frequency, f, incident at some point of interest. Assume that each independently radiated emission is represented as a simple cosine function. Let the ith independently radiated emission be written as:

$$E_i(t) = E_{0i}\cos(\alpha_i \pm \omega t)$$

where:

$E_i(t)$ is the radiated RF field level due to the ith independently radiated emission at time, t,

E_{0i} is the amplitude of the ith independently radiated emission,

α_i is the phase delay with respect to some agreed reference of the *i*th independently radiated emission,

and:

$$\omega = 2\pi f$$

The sum of these emissions at the point of interest can also be expressed as a simple cosine function at the same frequency:

$$E_0(t) = E_0 \cos(\alpha \pm \omega t)$$

where:

$E_0(t)$ is the radiated RF field level of the sum field at time, t,

E_0 is the amplitude of the sum field,

α is the phase delay with respect to some agreed reference of the sum field,

and:

$$E_0^2 = \sum_{i=1}^{N} E_{0i}^2 + 2 \sum_{j>i}^{N} \sum_{i=1}^{N} E_{0i} E_{0j} \cos(\alpha_i - \alpha_j)$$

Careful examination of this equation indicates that, to know the cumulative radiated emission, two pieces of information are required for each independently radiated emission:

- the amplitude, E_i;
- the phase, α_i, with respect to some reference.

While an SI will generally have knowledge of the amplitude, it will not generally have knowledge of the phase; it will therefore have only half the information required to use this equation.

The equation does allow the SI to calculate the extremes of emission performance. The worse case scenario is generated when all of the individually radiated emissions arrive at the measurement point in phase with one another and hence undergo constructive interference. In this instance, the amplitude of the system-level emissions, E_0, becomes:

$$E_0 = \sum_{i=1}^{N} E_{0i}$$

i.e. the simple arithmetic sum of the amplitudes to the individually radiated emission levels.

If it is assumed that the system emission level, E_{system}, is the absolute worse case combined field generated by some number, n, of identical items of apparatus, each independently radiating some level, $E_{apparatus}$, at some common frequency, then it is possible to write:

$$E_{system} = E_{apparatus} + 20\log_{10}\{n\}$$

To prevent the system emission level exceeding a limit, E_{limit}, it is therefore necessary to restrict the apparatus emission level to below the limit by the $20\log_{10}\{n\}$ margin, that is:

$$E_{apparatus} = E_{limit} - 20\log_{10}\{n\}$$

If the SI seeks to constrain the emissions of the system to the same limit as the constituent items of apparatus, adoption of this approach would lead the SI to define a new, tighter emission limit as part of its procurement process. As the number of items of apparatus within the system increases, the restricted emission level is unlikely to be met by COTS apparatus. As a result, the integrator is likely to have to organise the customised re-engineering of a COTS product. This will generally cause both significant delay to system deployment and significant increase to the unit costs (a result of reduced economies of scale in manufacturing and the need to recoup redevelopment cost). Fundamentally, this approach may lead to the specification of an emissions requirement that cannot be achieved.

6.3.2   System Emission Level Prediction — BT's Approach

Basing system emissions certification upon the absolute worst case is fundamentally flawed, since it takes no account of the likelihood with which this may occur.

The worst case approach assumes that the individually radiated apparatus emissions are in phase at the measurement point and hence interfere constructively. The phase relationships between the individual apparatus emissions are both unknown and uncontrolled — in principle, any value is possible. Given the *a priori* lack of specific knowledge, the individual phase terms can be assumed to be random — capable and equally likely to adopt any of the possible values (i.e. between 0 and 2π radians).

A simulation tool was developed to capture the separate apparatus emission levels generated at some measurement point and then generate both the probability and cumulative probability distributions that describe the corresponding system emission level. Compliance of the system is demonstrated through use of the cumulative probability distribution and the 80-80 rule. Use of the technique is best illustrated through consideration of an example.

6.3.3   Example

Consider the deployed configuration that consists of ten identical items of apparatus, each with a highest emission level of 25 dBµV/m at the common frequency of 100 MHz. All such items of apparatus would be considered as extremely well engineered, complying with the CISPR 22 Class A emissions limit of 40 dBμV/m (at 10 metres) by a margin of 15 dB.

The worst case system emission level is in this case $(25 + 20 \log_{10}\{10\})$ 45 dBμV/m, i.e. 5 dB in excess of the limit. To constrain the worst case system emission level to not exceed the CISPR 22 Class A limit would therefore lead the SI to re-engineer or seek alternate apparatus with a highest emission level no more than $(40 - 20 \log_{10}\{10\})$ 20 dBμV/m.

Figure 6.1 displays the probability distribution obtained for this example from the simulation tool. This displays the probability (or 'relative frequency of occurrence') with which the system emission level will adopt any of the possible values between the worst case (i.e. highest level) and the best case (i.e. lowest level). Note the general form: the distribution spans the range of possible values, displaying minima at the boundaries (i.e. worst and best cases) and a maximum value (this being the statistical expectation for the system emission level) somewhere in between. Hence the worst case system emission level can occur (it is indeed physically possible), but the probability distribution indicates that it is relatively unlikely to do so.

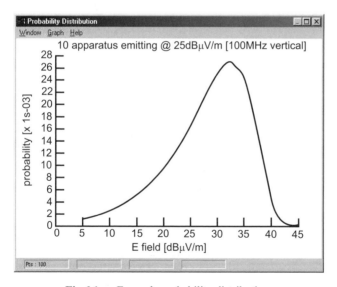

Fig 6.1 Example probability distribution.

The probability distribution can be applied in two ways. Since it displays the relative likelihood that a given system emission level will occur, it indicates the probability with which a given emission level will be measured. The distribution can be applied to emission levels generated by the entire population of deployment configuration. It can also be applied to describe the emission level of a single deployment over time.

Figure 6.1 indicates that for this example, the statistical expectation for the system emission level is 33 dBμV/m, i.e. only 8 dB above the apparatus emission level (25 dBμV/m) and 7 dB below the limit.

Figure 6.2 presents the cumulative probability distribution obtained for this example from the simulation tool. The cumulative probability is simply the integral of the probability distribution, displaying the statistical confidence with which the system emission level will be equal to or less than a specified value. It is this distribution that is used to certify the system using the 80-80 rule — as long as the system emission level corresponding with the 80% statistical confidence level is less than or equal to the CISPR 22 Class A limit, the system is compliant.

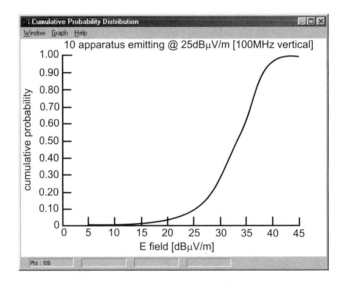

Fig 6.2   Example cumulative probability distribution.

Figure 6.2 indicates that, for this example, the system emission level will comply with the CISPR 22 Class A limit with 96% confidence — more than the required 80% confidence.

This example has served to demonstrate that while the system emission level **can** exceed the limit (by up to 5 dB in the worse case), there is only a 4% chance that this is likely to occur in practice. The statistical expectation for the system emission level is instead 33 dBμV/m (i.e. 7 dB below the limit level) and the statistical confidence in the population of this deployment configuration meeting the limit is 96%. Hence BT would present these results within a TCF to certify the radiated emissions of the system. It would also recommend that this configuration be deployed without any remedial engineering whatsoever.

This example also shows that it is possible to deploy systems composed from items of COTS apparatus, individually compliant with the radiated emissions requirements of the EMC Directive, and still certify the system emission levels with the same limit.

Note that while this example considers a system containing only identical pieces of apparatus, this is for the sake of simplicity — the approach is equally applicable to systems containing non-identical pieces of apparatus.

BT has filed a patent application in respect of this strategy [2].

6.3.4 Risk Management

In addition to supporting system certification, the technique also allows the risk of interference to be assessed for a given deployed configuration through a remarkably simple two-step process.

6.3.4.1 Step One

This typically takes place during the initial planning of a new system. At this point, detailed information on the apparatus emission performance is not available (either the procurement process has not identified the preferred apparatus or the preferred apparatus has not yet completed its development cycle). Details of the number of items of apparatus within each planned configuration are captured, along with estimates of the 'typical' separation distances between the deployment and a potential 'victim' radio receiver and a 'typical' estimate for any building attenuation arising when the systems' emissions leave the telecommunications building. The assessment is performed by assuming that all items of apparatus emit at the appropriate class limit of CISPR 22 at all frequencies. This assumed apparatus emission level is adjusted to account for the propagation through the building and to the 'victim' antenna.

The emission levels at this point are then subjected to statistical analysis to determine the risk that the system emission level will reach a point that could cause interference with the 'victim' receiver.

In this way, many scenarios can rapidly be assessed for different building attenuation values and different separation distances. If the derived risk of interference is deemed acceptable, no further analysis is taken. If, however, the risk of interference is deemed unacceptable, the second step is performed.

6.3.4.2 Step Two

An assumption that all items of apparatus emit at the limit line, at all frequencies, is a scenario likely to overestimate the risk of interference. If this is the case, a second step is performed.

In this step, the actual emission performance (i.e. frequencies and levels) of the apparatus is considered within the analysis. This generally reduces the total number of independent emission sources considered within the analysis and lowers the emission level considered. The same scenarios considered during the first step are reanalysed to determine the risk of interference.

6.3.5   Confirmation of Approach

The statistical distributions generated by BT's approach are supported within a TCF with the distributions obtained from a limited sample of emission measurements performed through *in situ* surveys. These surveys were originally commissioned as a means of validating the predictions of the simulation tool. To date, all measurements have served to support the statistical distributions generated by the tool: measured levels cluster around the derived statistical expectation for the system emission level rather than near the absolute worst case level.

6.4   Summary

The inclusion of systems within the scope of the European EMC Directive imposes a series of mandatory requirements on the systems that underpin BT's telecommunications network. Radiated emissions are of particular concern, since these are the most likely means by which a system may generate an interference complaint and hence risk disruption of the revenue stream.

The EMC Directive offers the TCF as the only practical and cost-effective route to demonstrating the compliance of large systems. Given the number and variety with which a given system is deployed, a statistical-based sampling approach to demonstrating compliance is demanded. The CISPR 22 80-80 rule offers an ideal solution to this problem. However, none of the conventional options for obtaining the required sample of emission measurements (consisting of *in situ* surveys, site-based testing and mathematical modelling) are cost-effective.

BT therefore developed an innovative strategy that adopted a risk-management approach. The statistical properties of a system's radiated emission performance are predicted from knowledge of the radiated emission performance of the constituent items of apparatus and the assumption of random phase between the apparatus emissions. The statistical description can then be used to certify the system emission performance using the 80-80 rule to the same emission limit as its constituent items of COTS apparatus. This strategy has delivered dramatic reductions in both the cost and time to declaring compliance. The approach also allows the risk of interference to be quantified for a given deployed configuration.

While the approach has been applied within the telecommunications and ITE domain, the method is readily applicable by any SI of a large system. This strategy is of topical interest given the likely inclusion of fixed installations within the scope of the revised EMC Directive that is currently under preparation — no other approach is perceived as delivering compliance with the combination of low costs and quantified statistical confidence.

References

1   International Special Committee on Radio Interference (CISPR) Publication 22: Information technology equipment — Radio disturbance characteristics — Limits and method of measurement (1993).

2   International Patent Co-operation Treaty (PCT) patent application No WO00/11483 (March 2000).

7

PERFORMANCE MODELLING

N W Macfadyen

7.1   Introduction

7.1.1   What is Performance Modelling?

What is performance, and why is it that performance modelling is different in principle from other types of modelling work?

In almost all other areas of telecommunications, modelling is devoted to determining the equilibrium mean behaviour of a system. Random fluctuations and perturbations are distracting factors which possibly must be allowed for, but that is done by robustness studies, or a series of planned variations to see the impact that they have around some fixed point. If possible, it is preferable to eliminate their effects at the outset. For performance, by contrast, it is the randomness that is the core feature to be studied: the entire area is devoted to examining the quantitative effect that randomness of input or environment has upon a system or network.

This means that performance models need a degree of accuracy which is unique among modelling techniques. Not only must they represent with considerable accuracy the target system being studied, but, in order to quantify what are essentially second-order effects, they must also take one of two courses:

- either these effects must be directly (i.e. analytically) calculable from some more robust base-level quantities;

- or the model must be so faithful that significant differences in output values can be confidently ascribed to the effects we are studying, and not instead to mere model inaccuracy.

The first choice above leads to the great category of analytic modelling techniques, of which the primal example is that of representing congestion on a circuit-group through Erlang's formula — which is so well-known and simple that it hardly any longer ranks as modelling. The second choice leads to the class of simulation models; where a computer representation of a system is built, and run until output is deemed sufficiently stable. We shall consider this further in due course.

In either case, however, there is a stern requirement on the modeller. While customers for studies may be saying 'model, that ye may understand', every proficient non-trivial performance modeller's mantra is rather 'understand, that ye may model'. Without a sound knowledge of the target system and its behaviour — and in particular the underlying reasons and mechanisms for that, and the uses to which the results are going to be put — little useful will be done. An obvious corollary is that the skilled modeller recognises the limitations of a model, and when it has broken down.

7.1.2   Why is Performance Needed?

We start by discussing why it is that performance understanding is of real importance.

So consider a simple case of a single group of (say) 120 classical telephony circuits. A static argument predicts that this could carry 120 simultaneous connections; taking into account randomness, however, will downgrade this to a mean value of just 103, if we want to offer a grade of service (GoS) so that only 1% of calls encounter congestion and have to be rejected (a typical order of magnitude specified for large networks). If we aim for a GoS of 0.1%, as is common on access links, that reduces to 93 Erlangs mean traffic; while imposing in addition the overload criteria in general use in BT, to allow for day-to-day fluctuations of traffic which we know to occur, reduces that still further, to around 84 Erlangs.

So the requirement of some quality of service has reduced the capacity from 120 down to 84 Erlangs. This is a very significant amount. The numbers are typical — they are not unusual extremes — and illustrate clearly that quality of service can be the arch-enemy of profitability. In a queuing system, where demands can wait for service rather than suffer immediate loss, it is even more important.

While a loss system such as the PSTN can be run at modest overload without real problems, that is by no means true of a delay system — which in practice encompasses most of the processing systems within the network. Offered a load even modestly above unity, a processor system has no equilibrium point: delays and queue sizes become unbounded in theory, and in practice the system collapses. Obviously the traffic must be sufficiently low that this does not happen; and controls are needed to prevent it during unexpected surges of traffic which approach this point.

These two examples, simple though they are, illustrate the two great issues involved in performance, which have no analogue in a static (non-random) environment: evaluating the capacity at a defined quality of service level, and defining controls which will preserve network operation and integrity when the unexpected happens.

Both of these issues bind up the detailed system behaviour with the randomness of the environment and its drivers (the traffic) in an essential way. Performance modelling attempts to understand this and its consequences.

7.1.3   Why Do We Model?

Performance modelling is open to the obvious challenge: 'Why should a system be modelled, as opposed to being straightforwardly measured in the laboratory?' For many other areas of modelling, this question does not arise — it would be considered absurd, for instance, in the context of business modelling — but it is a serious challenge and deserves serious consideration. While for many systems this option may indeed exist, there are clearly a number of areas where it does not. In no particular order, these include the situations where the target system is:

- too large — a global Internet cannot be assembled in the laboratory, nor can comprehensive sets of measurements be taken without great time and expense;

- too fast — equipment to load and measure a system must obviously be significantly faster than the system itself;

- too new — monitoring equipment may not yet exist, or, alternatively, the target system may be in development, with the aim of the modelling being to influence its design prior to production;

- too detailed — it may not be possible to monitor internal system quantities (e.g. software process queue-lengths or occupancies);

- too critical — intrusion upon business-critical networks or secure defence systems is generally deprecated;

- too speculative — an assessment is needed of response to conditions which do not yet exist, such as after traffic growth or with a different traffic mix;

- too process — the aim is to study the effects of different possible process design or management activities, which by their nature are long-term and in any case cannot be altered without great upheaval.

In all these cases, and others, direct measurement is not a realistic possibility, and so, for a full assessment, actual performance modelling becomes necessary. Where the target system does exist in some form, this may be augmented or validated by comparison with particular measurements as and when they are available; but to produce useful advice and guidance for the target system's owners and operators, such spot measurements are of very limited use by themselves except as a reassurance upon current performance, with current traffics.

7.1.4   Uses of Models

Performance models are required for a great range of uses. However, the performance life cycle is, to a considerable extent, the direct complement of the standard profile of attention devoted to a system.

At the exploratory stage of a new technology, performance modelling is always required, to demonstrate its raw capacity and capabilities, and to identify any issues of stability or control. (An outstanding example of this is the effort and success of the performance modelling of ATM generally, and of ABR in particular.)

During the launch phase of a product, however, traffic is generally far below the capacity of the minimum possible system, and all development effort is devoted to achieving a fully functioning network. There is frequently little spare budget, and no demand for performance studies. It is only after the system is in place, and initial growth has subsided, that the need for the system to match demand and capacity, and to make money, reinstates a need for the sophistication of performance modelling. To a considerable extent, therefore, performance modelling is by its very nature always out-of-phase with the current immediate preoccupations of a high-technology business.

There is, however, another issue involved — that of the stability of input requirements. While in principle it may be possible to say that the capacity of a system that just satisfies a given performance requirement is (say) 100 units, in a real operating context the demand will fluctuate unpredictably from one day to another, and the decision is likely to be taken, to plan to upgrade as soon as the mean level reaches (say) 70 units.

This is associated with the shape of real-life response curves. Naïve expectation — generally not consciously expressed or acknowledged — is frequently that performance characteristics degrade with load gracefully, as in the dashed line in the hypothetical Fig 7.1; in which case, it makes perfect sense to seek to run at a mean congestion level of 10 and load of 60% since the inevitable fluctuations in load will have perfectly acceptable performance consequences.

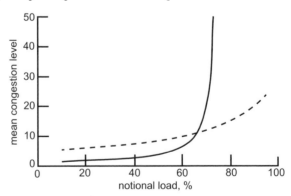

Fig 7.1   Idealised performance characteristics.

Real systems, however, frequently behave qualitatively more like the solid curve; in which case, the sudden take-off of congestion means that, even if the nominal target congestion measure is 10, it is not sensible to attempt to run at that level: rather, every process is tailored to ensure that the system runs well below this level in order that random daily variations in load never cause the extreme congestion that even minor upwards fluctuations can do.

Performance studies in this case therefore have a different type of use: they may not be a map specifying the precise path to tread, but they do reveal the existence and location of a precipice to be avoided, so that a decision to operate at (say) 50% load instead of 60% can be taken on an informed basis. In any case, while there is always excitement when strange pathologies are found, a reassurance that the performance of a system is smooth and gently varying is itself, in fact, a valuable and important contribution.

The illustration here is trivial, but more interesting real-life examples abound. Besides simple monotonic characteristics like those above, performance modelling has revealed the possibilities of network-wide instabilities allowing even classical voice networks to 'flip into congestion', if widespread AAR (automatic alternative routing) is employed; hysteresis loops in random-access systems with distributed queues; instabilities in IP networks with particular operating conditions; and many others. Of these systems, some have disappeared completely (e.g. the Cambridge Ring) once their peculiarities were established; others have had protocols modified to ensure stability; while yet others (e.g. the PSTN) are operated in regions where the risk of catastrophe is negligible.

Performance models therefore span a wide range of complexity and sophistication. Here we give examples drawn from the various extremes — of a simple spreadsheet-based model of voice packet jitter over an ATM network on the one hand, through to a very large and sophisticated model of the entire PSTN on the other.

The effort necessary for these differs by several orders of magnitude: while component elements can be rapid, full network modelling studies can be sophisticated and costly, justifiable only for large and stable networks.

The general tenor of this chapter is oriented towards the performance of networks at transport level, as opposed to the performance of system or process, but that reflects merely the examples chosen. The problems encountered, and the techniques used to solve them, are common to them all.

7.1.5   What Performance Modelling Is Not

Finally, performance modelling is not about design, although it is clearly closely related. Designing an optimal (in the sense of least-cost) network for a customer is an important and always highly relevant problem which certainly falls under the general heading of performance, but not of performance modelling.

There is indeed no direct route from performance to design — a discontinuity which is not always understood by non-practitioners. There is no straightforward process which allows the systematic design of a network to satisfy given performance criteria: the only process available is an iterative one of design using guidelines which are more or less refined, followed by evaluation of the output to see whether it meets the performance criteria that were specified.

We shall not therefore consider the issue of design and planning tools any further.

7.2 Modelling Techniques

7.2.1 Analysis

The most valuable technique available to performance modellers is that of mathematical analysis. This is not perhaps always regarded as modelling by other fraternities, although it is difficult to see quite why: its very essence is that of understanding the behaviour of a system, and constructing a mathematical representation thereof in order to quantify its behaviour.

The title does not imply that an explicit analytic equation can be written down subsuming everything that is needed. Such is indeed exceedingly rare, and confined to rather simple situations. Even when it can be done, the expressions concerned are generally opaque, and need numerical evaluation before very much can be deduced from them. For instance, an expression built around the roots of an equation with transcendental coefficients, or one based on the eigenvalues implicit in a large matrix, has a very low perspicuity rating. What the term 'analytic' does imply is that no simulation and no empirical input is employed, apart possibly from descriptions of traffic or other exogenous factors inaccessible to the modeller. The actual implementation may involve significant computation, although now a gratifying amount can be done by spreadsheet techniques.

Despite the increasing power of simulation, analysis is still the preferred option. It allows rapid evaluation of a range of scenarios, can be incorporated as a sub-model into larger constructs, and is a simple and elegant way of downstreaming results to others. It generally provides a deeper understanding than does any alternative procedure.

It has, however, obvious limitations. Chief of these is the range of systems it can study, and the fact that changing something qualitatively can be very difficult or impossible (for instance, a simple change in queue discipline can turn a model which is simple into one which is totally intractable). An associated danger (albeit one that arises from its very flexibility) is that of inadvertently overstepping the regions of parameter-values for which the operation of the system itself is properly understood.

This danger often arises when some component has been represented statistically: statements in academic texts such as 'the decrease in faithfulness of the model is more than compensated for by the increased facility of the analysis' are invitations to misuse. When a distribution is built in to a model, it is easy to misapply it by passing to a region where it does not make physical sense: representing a number of items in a system by a normal distribution may be appropriate when the range is [0, 1000] and the mean is 500, but will lead to very strange results if the range is [0, 10] and the mean is 2. The same applies to innocent-seeming boundary conditions: in conjunction with the very rapid decay of the tail of the Gaussian, this has the potential to change results not by a few percent, but by many orders of magnitude.

In general, therefore, analysis is the universally preferred methodology, but tends to be restricted in its application to the more simple, relatively straightforward problems, where there is little need to study the comparative effects of different qualitative changes to system design. It is not always the difficulty in modelling that is the constraining factor: even experienced modellers may not be able to give a categorical advance assurance of the complete accuracy of their analytic models, and so it is often necessary to construct a simulation, not because an alternative analytic model is not identifiable, but simply to reassure and convince others.

It is a sad truth that simulation often convinces more than does analysis, sometimes because mathematics is regarded as inherently suspect (because incomprehensible), and sometimes because the degree of abstraction necessary to produce an analytic model gives rise to uneasiness.

7.2.2   Simulation

Simulation is therefore the general (though not universal) last-resort method for performance modelling. Constructing such models is time-consuming — the representation of those parts of the system of no interest other than as background can take more effort than all the rest — and absorbs further effort in testing and validation. And when complete, a series of lengthy runs must be made in order to obtain a single one-dimensional view of the problem-space.

Tools and techniques for this have, however, evolved rapidly over the years, and rendered such models easier to construct and more descriptive (and, it is to be hoped, less error-prone). Between them, they have drastically reduced the time to construct complex models. Singleton [1] illustrates the ease with which a typical such package can be used to model an operational system.

The increase in power of such tools is substantial indeed. These have evolved from the mere provision of helpful modules in general-purpose languages; through special-purpose object-oriented languages such as SIMULA, which assisted the mechanics but were otherwise no great advance — through special-purpose packages such as BONeS^{TM}, to a large number of present-day proprietary packages widely used by performance engineers and providing a wealth of functionality, which it would be invidious to mention individually. Present packages allow very extensive drag-and-drop construction of models; facilities for hand-crafted interface code, replacing modules by new ones, and visual displays of almost any perspective of a simulation run's progress; extensive libraries of component models of important subsystems and processes (such as ATM links or TCP/IP); and powerful facilities for interaction and debugging.

Some network-modelling packages now even include functions for automated network discovery and traffic measurements, in order to facilitate the building of overall network models. It is important for the performance modeller not to be seduced by the lure of such seamless functionality: the automated selection of a default model or algorithm is no guarantee of its validity or reasonableness, and the real accuracy with which a simulation model can predict network quantities is still often below what is needed. Not surprisingly, the claims of package vendors should not be taken uncritically at face value.

The increased processing power of hardware has also rendered some of the major issues of previous years largely irrelevant. Memory constraints seldom exist any more, and the difficulties encountered in trying to fit large switching-system models into a maximum memory of 4 Mbytes are only a distant memory; and run speeds are such that warm-up time (provided the need for it is recognised at all) is no longer a real drawback.

Similarly, the ability to run a model for considerable periods, while viewing a range of output in graphical format, has rendered the concept of confidence intervals rather less of an issue, except for simulations which involve time-varying systems, the response to transients, or rare events.

7.2.2.1   Model Reusability

It has also made relevant once again the longstanding (and tantalising!) question of reusable models. The experience of recoding a model of an already-studied effect, because it is needed as a component part of a new model, has been widespread; and the corresponding question of whether the old model could not be reused has been raised constantly for many years. It seems wasteful virtually to duplicate work done previously.

The practical situation has been rather different. The focus of interest moves on. Performance modelling may start with the details of a particular low-level protocol; having established that it works, its next appearance is as one component in an entire protocol stack; later, it is needed only as a small component in the overall network model being written to establish traffic control mechanisms. The features of the protocol that need modelling change accordingly, with intimate detail being replaced by broader-brush impressionistic representations; and retaining the lower-level detail is not just a drain on the simulation run time, but implies major effort to code up the corresponding detail elsewhere for compatibility. It is often far easier to code up a dedicated new model.

New simulation packages are changing this. There are now large libraries of sub-models available, and using these is rapid and productive, but they are still not a universal panacea. It is still the case that levels of detail vary dramatically; and models are reusable only to the extent that the same or similar problems are being revisited again and again. But such packages have indeed succeeded to a considerable extent in producing that holy grail of performance — the genuinely re-usable module — and the speed of application-building has soared as a result.

7.2.3   Hybrid Models

A third class of performance model is the hybrid model. This represents a coming-together of the features of the analytic and simulation, possibly even with the addition of data or trace-driven input as well. The latter is particularly useful to study how different systems would react to a particular stream of input demand.

The general use of this category is to combine analytic submodels of component parts of a large system. We give an example of it in section 7.2.4 below.

7.2.4   Choice of Methodology

The choice between analysis and simulation is frequently clouded by feasibility.

A classic example of this is in the well-known case of Brockmeyer's equations for the distribution of traffic upon a finite overflow group of circuits. An exact closed-form analytic solution has long been known: unfortunately, however, it is notoriously difficult to compute since it is a multiple summation which, while finite, relies upon detailed cancellations between individual terms. In consequence, its calculation for more than a very small circuit-group size has been totally impracticable. On the other hand, a simulation is simple to write and very quick to run because of the inherent simplicity of the system.

Another illustration of this can be drawn from Carter et al [2]. In that paper, the effect of using for control purposes a moving-average estimate of a queue length, instead of its actual instantaneous value, was studied. Some systems, however, actually use, instead of the true 'moving average', a quantity which (broadly speaking) coincides with that for upward movement but is liable to discrete downward jumps as the queue length falls. Modelling that analytically is challenging, but can be done (though at the expense of some approximations). Unfortunately, the structures required are complex, and absorb much more computing effort than does relatively simple direct simulation — which has the added advantages of comprehensibility and lack of need for approximation!

The balance of practicability can, however, equally well flow in the other direction. A topical example of this arises from the study of voice over ATM (where another issue is considered in section 7.5 below). ATM switches have separate buffers for constant bit-rate (CBR) traffic, and it is usual to plan to run these at not more than 85% occupancy: but voice traffic is more regular, and the question arises of how far this can be extended while still keeping cell-loss rates acceptable. In this case, the choice is between an ND/D/1 analytic queuing model, using delay as a proxy for loss, and a simulation. The choice is weighted against the simulation because of the rareness of loss, which impinges upon the run-lengths required; but the deciding factor is the explosion in memory required by the simulation package because of the finite-source nature of the problem. The outcome is that while for small systems the simulation is swift and convincing, for more realistic system sizes it becomes impracticable, and the analytic model reigns uncontested. The only justification for simulation is therefore as a validation of the analytic model at small system sizes.

7.3   Validation of Models

The issue of model validation is a universal bugbear.

Where the target system is well-known and well-defined, comparison of model results with actual measurement data may be feasible; but this is definitely a rare occurrence. The more usual situation is one where either the target does not yet exist, or where the real system has so many uncertainties and complicating factors that detailed comparison is impossible.

Typical instances of the latter are given by the validation of proposed control mechanisms. Two of these were identified in a recent publication [3], and we shall consider them briefly here.

The first of these [4] studied a mechanism for protecting ordinary voice traffic in the network, in the presence of aggressive IP-dial traffic with different characteristics. No straightforward method exists of validating this model by comparison with data: in practice, not only is the variability of traffic huge, but its statistical characteristics at any instant are unknown and unknowable. Only the grossest comparisons can be made, basically to determine whether any mechanisms proposed seem to have the overall effects predicted. An added difficulty is that the model studies proposed mechanisms, not existing ones: until these have been agreed, accepted, and rolled out into the network on at least a limited trial basis, there can be no data to study. The apparent option of implementation upon a restricted test-bed is not useful, because the model being validated contains, as an integral part, a representation (i.e. model) of the different types of traffic involved: so no test-bed, which relies upon simulating through test call generators or otherwise, is a really adequate approach. It is important to ensure that a validation does indeed confirm the whole model, including the assumptions, and not just the mathematics involved.

The second model [5] proposed a sophisticated and adaptive technique for controlling call-attempt overload in the network. Very similar considerations apply here — there is no way in which comparison can be made, until at least full development and trial roll-out have been completed. And it is of course just such development and roll-out which the model is attempting to justify.

This is indeed a general difficulty with large network-wide models. Another prime example of this is given by the case-study presented below, of overall grade-of-service studies which modelled the interacting effects of all parts of the network, and their interaction with management processes. Short of extensive trials in which the whole operation of the entire network was disrupted for several years, no direct validation is possible.

The approach to validation generally adopted is therefore indirect. There are in fact at least four conceptually distinct areas to verify; in increasing order of difficulty and uncertainty, these are:

- that the software implementation operates correctly;

- that the system or network operation is correctly represented;

- that the model's simplifications and techniques are valid;

- that the model's underlying assumptions (e.g. of traffic or user behaviour) are valid.

The first of these represents no special difficulty, as it is basically just software testing; and the second is straightforward but often time-consuming, including lengthy discussion and review by system owners or designers. It is the third and fourth that represent real difficulty; and obtaining sufficient assurance that in-built models and representations are adequate is not easy. This is frequently so where statistical or large-number approximations have been used (e.g. representing a random walk by a diffusion process) to simplify models, when in fact the range of parameters involved raises questions as to the accuracy of these.

Building sufficient confidence as to overall model accuracy therefore can involve a whole series of minor studies or arguments as to the robustness of model output to different effects. A simple and classical example of this is the invariance of Erlang's formula for standard loss probability, to the precise form of the call holding-time distribution: a robustness which does not extend to waiting models (the M/M/1 queue is very different from the M/D/1). Modelling packet networks assuming negative exponential (negexp) service times may be convenient analytically, but it is important to understand that this is an approximation, and model validation will involve checking that the quantitative effects of this are acceptable. Similar consideration has to be given to any assumptions upon statistical forms of distributions, and (especially) to independence assumptions.

At this point it is worth mentioning (and deploring) the way in which critical assumptions can be buried in mathematics, and then accepted unquestioningly.

Reviewers and target audiences frequently do not have the insight or expertise to understand in detail quite what is being done, and impressive analysis can (as with many professions) engender a totally misplaced faith in the accuracy of the result. While straining at the gnat of data accuracy, it is all too easy to swallow the camel of model assumption.

Checking that the overall model is such that special simple cases (if any exist) reduce to known and standard results, which is really more akin to software testing than to the more intangible model validation, is therefore nonetheless an essential part of the validation process. While in a sense trivial, it does at least give reassurance that the model is not egregiously wrong everywhere.

7.4   Drivers to Models

Although it is possible to construct a pure simulation model using straightforward and well-defined techniques, to draw up an analytic model needs judgement. One particular area of relevance here is that of the model's drivers. In most other fields, input data is fixed, or varies only between scenarios: in performance, input data is by definition the set of random stimuli that drive the system, and the description of these has to be appropriate.

A tension exists here between a fuller and more elaborate description — which may describe the traffic well, but is likely to be both uncertain with regard to its parameter values and very probably too complex to make use of analytically — and a simpler one which is robust and straightforward in application, but may not adequately represent reality. An introduction to some of the ideas of traffic characterisation that are relevant here can be found in Macfadyen [6].

This specification of traffic drivers is of the highest importance for successful modelling, but it is also an area where experience and knowledge play a more important role than is strictly desirable. While in some cases this only affects the mathematics (trading off fidelity of description for tractability in analysis), in others it can quite fundamentally affect the whole behaviour of the system. Analytic techniques for dealing with this have ranged from simple models which adjust only the magnitude of the offered traffic (keeping unchanged, that is, statistical characteristics other than the mean), through passing to richer probability distributions (such as hyperexponential for overflow traffic, or Markov-modulated Poisson processes), or the introduction of highly complex analytic structures to take explicit account of correlations and the non-Markovian nature of the process, or their representation through generic processes whose statistical characteristics can be modelled, but whose details do not correspond to anything specific. Examples of the last include self-similar traffic streams characterised by a single (Hurst) parameter [7], and effective-bandwidth methodologies in ATM [8]. While the mathematics is frequently powerful and elegant, the experienced performance modeller treats all such models with caution, because of the gross simplifications involved at this stage.

Even when the situation is apparently quite simple, there is frequently a choice which must be made; and it is part of the modeller's task to do this appropriately, while assessing the effects and (if significant) making clear the risks for the conclusions.

An example of this can be found in the variety of mathematical treatments of ATM cell delay or effective bandwidth, predicated upon different underlying assumptions as to the cell arrival processes [9-13]. Choosing between these requires more than a superficial understanding.

An area where this is of particular importance is where the number of real sources of traffic in a system is finite, as opposed to essentially unlimited. Detailed consideration has to be given to their own behaviour, and especially so under conditions of congestion.

To illustrate this, consider a simple system of a group of N circuits being offered random traffic from S sources — a very elementary system, which as $S \to \infty$ becomes the classical Erlang system. The difficulty is about the behaviour of the user when he encounters congestion and there is considerable lack of clarity both about this and about the definition of the 'offered traffic'. The analysis all hinges upon the concept of the 'when-free calling-rate' of a source. Without much difficulty, we can define at least the following reasonably plausible models of user behaviour:

- fixed pre-specified when-free calling-rate;

- repeat attempts — a constant offered traffic, with a perseverance r on meeting congestion (experimental studies have found r to be around 0.85 in telephony networks);

- constant demand rate — demands arise at every source at a steady rate at all times, and they are carried as follow-on calls if they arrive during a source-busy period, but are lost if otherwise blocked because all circuits are busy.

None of these models has a uniquely convincing claim to superiority, and indeed situations can be envisaged where any one of them is relevant: the one to choose depends crucially upon the operating scenario to be studied. Their effects on a system are, however, very different.

Figure 7.2 illustrates the differences between these models, by plotting the single-attempt blocking probability for a variety of them, in a system with 20 sources and 8 circuits — the base case of constant when-free calling-rate; the classical Engset system, which is the case of fixed offered traffic but with zero repeat attempts; the cases of perseverance $r = 0.5$, 0.85 and unity; and the case of semi-queuing or constant-demand.

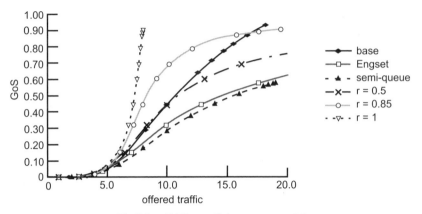

Fig 7.2   Different finite-source models.

The differences are clearly major: indeed, the effect of changing the offered traffic pales into insignificance when compared with that of changing the traffic model concerned. While at low blockings they all coincide (as they must do), as congestion rises they show substantial differences. An effect of the model which is at first sight paradoxical is that the 'offered traffic' generated can exceed one Erlang per source; and this, while perfectly correct in its own terms, warns clearly that the user-behaviour assumptions require very careful examination.

Because of this, modelling which treats of systems in congestion has to be especially careful about the description of the sources — or, because that is what is then generally the issue of relevance, the behaviour of proposed control mechanisms. The need to specify behaviour from a user which is both realistic, tractable, and convincing, while also being able to be varied across a wide range of different scenarios, frequently implies that simulation is the only methodology that can be convincingly used.

7.5   Analytic Case Study — Voice Over ATM

As a first example of performance modelling, we consider a particular small system which is ideally suited to analytic modelling: the jitter experienced by voice traffic when carried over ATM.

At the heart of BT's new next generation switch (NGS) voice network, which forms the core of the public switched telephone network in the UK, lies an ATM switch, the Ericsson AMD301. Within this core, voice traffic is both carried and switched as ATM rather than as time division multiplexed (TDM).

For sound operational reasons, and for compatibility with existing TDM networks, the basic unit of modularity is retained as a virtual path (VP) supporting 30 calls and one signalling channel. The individual voice paths are carried on bearers which are presented with the AAL1 adaptation layer protocol and so appear to the ATM platform as CBR virtual circuits (VCs) within the VP. Silence suppression within calls does not take place. We note that the VCs on the VP do not have the invariant clocked structure that the individual TDM time-slots do in their frame.

VCs which are not currently carrying (possibly silent) voice are identified in the NGS, and are then suppressed by the ATM switch. This is critical to what follows: the alternative scenario, of carrying and switching idle cells carefully across the entire network, gives rise to no problems.

7.5.1   Delay Causes and Requirements

When calls are carried by ATM rather than TDM, a number of new possible sources of delay exist, including:

- cell fill delay — the time spent to assemble sufficient data to justify sending an ATM cell;

- network-induced cell delay variation (CDV);

- shaping delays;

- play-out delay.

These build on the existing TDM sources of delay, arising from digitisation and coding, transmission, and actual switching. Further delay would be introduced by voice activity detection (to enable silence suppression); or by more sophisticated coding schemes than the standard G.720 if these were used.

A particular issue of interest is the size of the play-out buffer that must be supplied to allow for jitter (i.e. the random fluctuations that the arriving cells show about their nominal plesiochronous scheduled arrival instants) that is introduced by the network. Jitter requires a trade-off between introducing extra delay (and network cost) on the one hand, and the acceptance of dropped speech packets and degraded voice quality on the other. This is an ideal situation for analytic modelling.

A major cause of the jitter of individual voice cells in this system is the set-up and clear-down of fresh calls. Upon initial set-up, a call is assigned a position within the stream of VCs multiplexed together on the ATM switch egress port; but due to the comings and goings of other calls, its relative position within the set of calls carried in the same VP is subject to change. If the VCs are shaped on to the VP — that is, if they are not transmitted as they arise at the switch, but an attempt is made to smooth them to the nominal data-rate of the VP — this change of position can result in a very significant change in network delay.

Figure 7.3 illustrates this. For the sake of clarity, we have illustrated this as if the VP had fixed slots. A total of nine TDM cells arrive to an empty VP (i), and are played out after shaping on to the VP (ii); there is a single empty slot after cell 7, because cell 8 is not yet ready by this time. After several frames of this, a new call set-up arrives while cell 4 is being played out: this is after the arrival of cell 5 but marginally earlier than the existing cell 6, and so the new cell N pushes in before that cell (iii), so that cells 6 and 7 both slip upwards by one slot. Cells 8 and 9 are unaffected. When in due course the calls presenting cells 2 and 3 clear down, cells 4 to 7 all slip down one cell (iv), but the rest are unaffected because they are already at their earliest position.

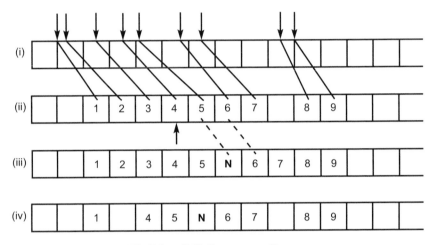

Fig 7.3   Cell slippage on call set-up.

It is clear therefore that shaping will introduce significant jitter, and hence the need for appropriate play-out buffers. Without any shaping, on the other hand, the potential is very high for cells to be discarded by a too strict ATM network policer as being in breach of the relevant cell delay variation tolerance (CDVT). In Fig 7.3, the cells labelled 2 are clearly candidates for rejection by such a policer if set too low, which would of course effectively block transmission in that direction on that call. While the effects of this would be limited to that single call, the consequences if an SS7 signalling link (also carried over such a VC) were to lose cells in this way would be far more serious.

7.5.2   Analytic Model

Consider the placing of individual calls on to a 2 Mbit/s VP. Assume as a worst-case that every VC in the VP is present, i.e. we are studying a single 30-channel system in a busy period. There is then no more than a single empty slot (just after a call cleardown, and before the arrival of another).

At the cleardown itself, there is therefore the possibility that some of the established cells will be slipped downwards by one slot in the VP; and at the next call arrival, there is an opportunity for some existing calls to be slipped up. The detailed probabilities depend on both the relative order and position of the calls (to establish the frequencies of slippage), and on their absolute location within their own TDM frames (to set the floor for downwards slip). A simple model, however, is easy to produce. Taking into account the fact that in this model (of a busy circuit-group) set-ups and clear-downs are not independent, but occur in matched pairs, we suppose that after such an event-pair a single call which is unrestricted by boundary conditions — which is the most slippery case — can have the outcomes of a single upwards shift (with probability ¼); a single downward similarly; and of no change at all, with probability ½.

By introducing the concept of an excursion, i.e. the distance that a given cell has moved from its nominal position, the excursion probabilities $P_k(.)$ after k event-pairs of clear-downs/set-ups are restricted to the domain $[-D, 29-D]$, where D was the cell's initial random delay within the frame; and away from these boundaries satisfy:

$$P_{k+1}(n) = ¼\,P_k(n-1) + ½P_k(n) + ¼P_k(n+1)$$

with special conditions at the ends of the range. In order to calculate the probability of the maximum excursion in a specific period reaching a particular value D_{max}, say, we need only introduce absorbing barriers in the probability space at this distance from the starting-point.

This whole process is both conceptually simple and computationally quite straightforward. Figure 7.4 shows how these probabilities vary with excursion length, for various fixed-call durations (expressed in mean holding-times).

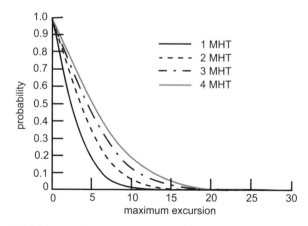

Fig 7.4   Maximum excursion probabilities for fixed duration.

Drawing the appropriate conclusions upon play-out buffer sizing, and upon the ATM CDVT required on the VP if shaping is employed, is then a simple application of this analysis; and the decision on the preferred operation of the network (in terms of shaping, policing, and CDVT) can then be made with complete confidence.

7.6   Hybrid Case Study — Overall Grade of Service

An excellent example of a large and successful hybrid model, where both analysis and simulation are used to support each other, is provided by the overall grade of service (OGoS) model constructed to study the actual behaviour of the PSTN as a whole, taking into account all known effects from networks, systems and processes. While this is perhaps atypically large, it illustrates well a range of techniques which are in widespread use on a smaller scale.

The aim of the study was to quantify the interaction of all the factors which affected the end-to-end grade of service (GoS) as seen by a customer, so as to obtain (for the first time) a real understanding of the achievable behaviour of the network as a whole. While individual circuit-groups are sized to particular nominal GoS levels, it is not immediately clear how directly these determine the overall perceived effect. A knowledge of this is important, not only for purposes of network operation, but also in order to achieve a balance between the conflicting demands of service quality, operational complexity, and cost.

Among the many contributing factors, all of which required detailed representation, we can include:

- network issues:
    - — network structure and topology;
    - — routing strategies;
    - — network faults;
    - — dimensioning methods and target GoS levels;
- traffic issues:
    - — traffic variability, including effects of overloads;
    - — growth;
- process issues:
    - — traffic measurement;
    - — forecasting and provisioning processes.

Representing the effect of all these in a meaningful way can only be done by presenting the model output in terms of the GoS distribution which is seen by a typical customer. Because of the multiplicity of factors in the list above, the GoS experienced by any one traffic stream on one day may be very different from that seen by another stream, or upon another day; and so it is only by assembling the distributions of all of these that we can attempt to get close to representing the service quality seen by a typical customer.

The obvious and natural approach to such modelling would be to construct a full detailed representation of the entire BT network. Initially, however, it was decided that a higher-level model was preferable with as much abstraction as possible, both because of the sheer size of such a simulation model and the need to produce fairly rapid results from the study.

The key observation to make about the PSTN which underpins this whole treatment, is that it is link-engineered, i.e. everything is determined on a link basis, and there is no attempt to treat any traffic stream on an end-to-end basis. In consequence of that, links can be treated separately and independently, and there is no need to take into account correlations along the length of any path or traffic flow. (Note that this is a major distinction from the type of modelling that is used for an IP network, as discussed, for instance, in van Eijl [14].)

Initial study of the problem served not only to establish appropriate division into sub-models, but also to verify that the statistical methodology adopted was valid. This point is non-trivial — assuming, for instance, product-forms for combining distributions, or for the distribution of traffic over a network, rest on implicit assumptions which need validating at as early a stage as possible because they will underpin the entire model construction. Since the assumptions stand or fall by how well they represent the data, it is by comparison with such data that they must be validated.

Construction of analytic models of, for example, traffic variability is straightforward, but obviously does require a knowledge of its statistical characteristics. Unfortunately, traffic variability from one day or week to another is itself highly variable and not reliably known, and subject to constant change as new services are introduced and network traffic changes. It is therefore crucial that such models be constructed with a suitable range of parameters, designed to cover all plausible ranges of behaviour, and that the final overall model runs should include a study of the sensitivity to these. Since a high sensitivity would mean that little confidence could be placed in the overall results, advance assessment of all these was essential.

The GoS for a given traffic stream is obviously determined by the routing that it sees in the network. The PSTN has an exceedingly rich set of alternative routing options available, including the proportional traffic distribution facility (PTDF), which operates so as to spread the load yet further. Incorporating all of these in the model decreed that the structure of the latter should separate its modelling of access and core.

The structure of the overall model was therefore defined as a collection of submodels, some of these (such as the traffic behaviour) contributing through providing input distributions for calculational purposes, and others through providing a component part of the end-to-end GoS distribution itself, needing appropriate combination with other parts to yield an overall figure.

Implementing this was not straightforward. While the core network itself was well understood and modelled, only an approximate analytic methodology was available for the access network. This represented the qualitative behaviour adequately, but suffered from an overall normalisation error; in addition, it was time-consuming to calculate. To overcome this, therefore, a hybrid procedure was adopted:

- calculate analytically the quantities required, on a (necessarily sparse) array within the up-to-five-dimensional space of interest;

- perform a single-point simulation and use the results of that to normalise the calculated points;

- estimate quantities at the (many) required points through interpolation of this normalised array.

Calculations of GoS distributions on the access network could then be made by using input distributions derived from the relevant sub-models discussed above; and these access models could then be combined through convolution-based methods with the core network functions. This technique, tuned dynamically by the up-front calculation of required cell sizes and accuracies, allowed the rapid calculation of a wide range of distributions of expected end-to-end GoS values.

The outstanding advantage of this structure for the model as a whole is its detail and speed. Analytic techniques allow a range of results to be produced which are impossible with simulation techniques — in particular, the calculation of actual distributions of GoS values, and precise and consistent figures for very small values.

We observe, however, that this is not a detailed model of the actual network: it was rather an exploratory tool, to enable the study of the effects of all factors, in a particular type of environment. The input data requirements for a single, straightforward, minimal run of the model required the setting of approximately:

- 9 discrete variables of yes/no or qualitative type;

- 23 parameter vectors describing continuous variables;

- 4 discrete model-tuning parameters;

- 6 continuous tuning parameters;

these being bare minima, with no comparisons or options selected; the precise counts increasing several-fold if multiple-element vectors are not counted as a single entity. These figures also exclude all the data necessary to describe the network route-size profiles which set the model in the context of any particular network (in the present case, BT's).

Although this hybrid model served its purpose well, after much use questions began to be asked which could not be accommodated within its structure. Some of these were concerned with detailed financial implications; but the critical questions were those concerned with the effects of dynamic alternative routing (DAR) upon the network performance. The effect of that is of course to give traffic streams an almost unfettered scope for overflow, and hence to couple together all the circuit-groups over the entire network — which destroys the critical assumptions that underpinned the model.

These can be overcome only by simulation; and for that reason the OGoS performance modelling moved on to the development of a full simulation model. For the purposes of this chapter, that does not offer the interest that the hybrid analytic model does; but it should be recorded that its scope, although much more detailed, was in a sense more restricted. We note too that all the sub-models of traffic behaviour, provisioning, etc, which were referred to earlier were usable again in the simulation context — only the actual network response required recoding in simulation terms.

7.7   Simulation Case Studies

A modern illustration, using an up-to-date package, may be found in Singleton [1], which typifies in many ways the type of short study that is often required. Lower-level examples of recent date would include studies of:

- the behaviour and response of various congestion-control mechanisms in the network [4, 5], including detailed study of their response to different user reaction to the mechanisms;

- the performance and interaction [2] of IP control mechanisms such as WRED, PQ-CBWFQ, WDRR, CAR, etc — the precise definition of these is not relevant here, only the fact that they represent rather low-level protocols, sometimes of extreme complexity, which require large, difficult and time-consuming simulations to model with any real confidence;

- the performance of ATM links carrying novel mixes of traffic, with particular requirements and characteristics;

- advance validation of throughput obtainable following major network technology changes;

- performance of different distributed processing systems or large database systems.

Such scenarios occur not infrequently, and form the core work of the simulation modeller. There is no longer much call for the detailed modelling of switching systems such as once formed the basis of much performance effort within BT; but it is of interest to recall briefly what was perhaps the most significant simulation model ever constructed within BT — the System X processor model.

This was originally built during the initial tripartite collaboration which designed and developed System X, and tracked that switch for many years. It was built entirely from a general-purpose language (PL/1), with a very few special-purpose routines for handling simulation-specific tasks such as random-number generation and event-list handling, and ran overnight on an IBM mainframe. Storage was strictly limited (both for the host machine and for the switch being modelled); and the complexity and run time required of the software meant that issues of warm-up time or confidence intervals were significant. No visualisation features at all were available.

Input to the model included data upon both the System X hardware and, of course, software. The latter included conceptual models of all significant tasks, together with detailed information on software process structure and run times, which were obtained from instruction counts and estimated branching probabilities of the software as it developed. Estimates of traffic mix were used to drive the model. Total development effort was substantial.

As with many performance models, little faith was placed initially in the results even by those who had commissioned the model — a cynical view may be that such studies are frequently an exercise to provide a defence in the event that the system performance turns out to be disastrous. It is gratifying to record that when the first real measurements became available from the switch, the performance modelling predictions were found to be accurate to within better than 1%: an outcome which raised the credibility of performance modelling, and the esteem in which it was held, by several orders of magnitude!

7.8   Summary

We have seen that performance modelling has characteristics of its own which make it uniquely deep and sophisticated. It has to represent not merely a system's static performance, but also its dynamic; and it also depends critically upon the faithfulness with which a wide range of intangible user-related factors can be modelled.

At the same time, it constitutes a rich field which has attracted study from industry and academia ever since its foundation. The problems it studies are frequently novel and with interesting mathematical ramifications (such as self-similarity or catastrophe theory); or they offer interesting real-life examples to validate statistical analyses upon convincingly large quantities of data which are just not available in other contexts.

Because of this, it also has a high mystification potential. It is all too easy for a report to obscure, through its apparent advanced mathematical rigour, the fact that a chosen model is only approximate, or that it really does not fit the usage scenario very well. The experienced and proficient performance engineer will draw attention to such drawbacks, rather than (possibly inadvertently) obscuring them; and takes care to present conclusions in terms that are relevant and simple to understand.

Despite all the progress with new tools and techniques, understanding and modelling a complex new system remains a difficult task, in which lack of experience can lead to conclusions which are highly misleading. A proper performance understanding of a large new system or network, by contrast, can lead to immense savings both in capital expenditure and in ongoing operational management effort. It is a crucial but regrettably unsung contributor to reconciling overall business profitability with customer satisfaction.

References

1 Singleton, P.: '*Performance modelling — what, why, when and how*', BT Technol J, **20**(3), pp 133-143 (July 2002).

2 Carter, S. F., Macfadyen, N. W., Martin, G. A. R. and Southgate, R. L.: '*Techniques for the study of QoS in IP networks*', BT Technol J, **20**(3), pp 100-115 (July 2002).

3 Ackerley, R. (Ed): '*Performance*', Special Edition, BT Technol J, **20**(3) (July 2002).

4 Stewart, N. M.: '*Potential interactions between IP-dial and voice traffic on the PSTN*', BT Technol J, **20**(3), pp 87-99 (July 2002).

5 Williams, P. M. and Whitehead, M. J.: '*Realising effective intelligent network overload controls*', BT Technol J, **20**(3), pp 55-75 (July 2002).

6 Macfadyen, N. W.: '*Traffic characterisation and modelling*', BT Technol J, **20**(3), pp 14-30 (July 2002).

7 Leland, W. E., Willinger, W., Taqqu, M. S. and Wilson, D. V.: '*On the self-similar nature of Ethernet traffic*', Sigcomm'93, **23**(4) (October 1993).

8 Kelly, F.: '*Notes on effective bandwidths*', in Kelly, F. P., Zachary, S. and Ziedins, I. B. (Eds): '*Stochastic networks: theory and applications*', Royal Statistical Society Lecture Notes Series, **4**, Oxford University Press, pp 141-168 (1996).

9 Anick, D., Mitra, D. and Sondhi, M.: '*Stochastic theory of a data handling system with multiple sources*', AT&T Bell System Technical J, **61**, (October 1982).

10 Gibbens, R. and Hunt, P.: '*Effective bandwidths for the multi-type UAS channel*', Cambridge University, Queueing Systems, **9**, pp 17—28 (1991).

11 Guerin, R., Ahmadi, H. and Naghshineh, M.: '*Equivalent capacity and its application to bandwidth allocation in high speed networks*', IEEE, JSAC, **9**(7) (September 1991).

12 Buffet, E. and Duffield, N.: '*Exponential upper bounds via Martingales for multiplexers with Markovian arrivals*', Dublin University, DIAS-STP-92-16 (1992).

13 Elwalid, A., Mitra, D. and Wentworth, R.: '*A new approach for allocating buffers and bandwidth to heterogeneous, regulated traffic in an ATM node*', IEEE, JSAC, **13**(6) (August 1995).

14 van Eijl, C.: '*Capacity planning for carrier-scale IP networks*', BT Technol J, **20**(3), pp 116-123 (July 2002).

8

COMMUNICATIONS NETWORK COST OPTIMISATION AND RETURN ON INVESTMENT MODELLING

A M Elvidge and J Martucci

8.1   Introduction

If we were not to model revenue and cost, it would be even more difficult to make investment decisions beneficial to shareholders. Such modelling gives an opportunity to run 'what if' scenarios before committing to a decision. Let's examine further.

It is no doubt due to the ubiquitous state-controlled telecommunications monopolies of the past that the tracking of costs in telecommunications companies has historically lagged behind other industries. In the car industry, the costs incurred in manufacturing a car are known to the smallest component. The cost base of a car is thus known to a high degree of accuracy, to within a few pounds. This kind of accuracy is the envy of the telecommunications industry and is a goal it is achieving.

It is not just the capital costs that need to be tracked. A car, once sold, becomes the owner's responsibility. The owner will need to maintain it, fill it full of petrol, service it and so on. The costs involved in running the car are no longer the responsibility of the car manufacturer. Networks, however, are invariably built and run by the telecommunications operator and so the responsibility for funding the cost of running the network is not transferred to a third party. In costing a network properly, the costs involved in running the network, the depreciation of the network assets, the utilities, accommodation, maintenance, people costs, etc, in short the operational costs of the network, need to be assessed and quantified. Only then can a telecommunications operator establish an accurate representation of the total costs involved in setting up and running their network.

Such accuracy is required, not only as a foundation for any business case that involves building or utilising a network, but also for regulatory requirements; a solid cost base ensures governing bodies can be sure the prices that operators charge for their products and services are reasonable and fair.

An accurate cost base will increase the confidence with which a wealth of financial checks and measures will be viewed. A project will have a certain net present value (NPV), payback, internal rate of return (IRR), cash flow and breakeven point. These are key indicators to the general health of a project and are more fully explained in Table 8.1.

Table 8.1   Project key indicators.

Indicator	Description
Cash flow	A project's income and expenses. Determines whether the project is 'living within its means' or having to borrow money each month.
Breakeven	The point at which a business is neither making money nor losing money, and managing to cover their costs exactly. This is a business' main objective from the start. Below the point of breakeven is a loss, and above it is a profit.
Payback period	Defines how long it will take to earn back the money that will be spent on a project.
Net present value (NPV)	Represents the present value, in today's terms, of the future net cash flow of a project. A negative NPV indicates a loss-making proposition. A zero NPV indicates that the project will just break even. A positive NPV indicates a real return on investment.
Internal rate of return (IRR)	Indicates the cost of capital that is equivalent to the returns expected from the project. Once the rate is known it can be compared with the rates that could be earned by investing the money in other projects or financial products.

An accurate cost base means that these measures provide an accurate reflection of the financial merits, or otherwise, of a project. Subsequently, the decision to invest in a particular product or service will be an informed one.

Costs relate not only to basic cost components, for example cost of fibre and duct, but also costs of basic cost components when used in the context of certain networks. This cost apportionment, getting costs in context, is key to an accurate cost base. The finance sections of BT have access to such costing information and are a key input to cost and business modelling work. Not only do they provide accurate cost information, but the costs come ready made with an audit trail, making validation of costs used in models straightforward.

Once accurate costs have been identified, the configuration of the network can be modified to find the optimised geographical combination of links and equipment that will lead to the minimum network cost. This cost optimisation approach could mean the difference between a project looking financially unappealing and one that suggests a healthy return on investment.

This general approach of identifying whole-life costs of a network (capital and operational costs), quantifying them via auditable financial sources and using optimisation tools to minimise costs, has led to more accurate financial models being built of potential network products and services. The increased accuracy has led to more aggressive pricing strategies, keeping BT competitive in areas where it may otherwise not have been able to compete.

8.2   A Structured Approach to Financial Modelling

BT's total cost of ownership (TCO) and return on investment (ROI) models generally address problems of the following type.

- What will a new network or service cost throughout its lifetime, in both capital and running costs?

- What profile of investment will it require?

- When will it pay back?

- How profitable will it be?

The answers are unique to the service or network, but the questions are similar, and so can be tackled with models that have a similar structure.

BT has adopted a structured approach to building financial models, based on a method originally developed for the City of London market [1], and adapted by BT for telecommunications analyses.

A reusable framework means that new models can readily be built, audited, and understood by developers, users, and customers alike. Microsoft Excel provides a familiar, friendly platform, suitable for handling large files of geographical data, when backed up with Visual Basic code for optimisation calculations. Models built according to this framework serve as communications tools, facilitating discussion with customers, and understanding of the issue on all sides.

The clarity of the model framework is partly achieved by segregating data and calculations into different parts of the model. Data is passed between worksheets, and complex linkages between the sheets are avoided by keeping calculations wholly within a single sheet. This makes for speed in testing, debugging and developing the model further.

Navigation of the model is facilitated through use of a 'Navigator' page which provides both an overview and logical grouping of worksheets within the model. Figure 8.1 shows the front page of a typical model. The 'Navigator' page contains buttons that take the user to worksheets that each have a very specific function.

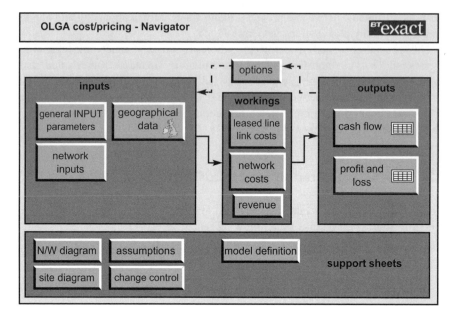

Fig 8.1   Model front sheet and user interface.

The model shown in Fig 8.1 is an example from a project codenamed Olga, which looked into different ways of providing Internet and corporate intranet access from hotel rooms, and assessed the potential costs of ownership of the infrastructure and the likely return on investment.

A particular feature of a typical model is the 'options' sheet, which gathers together the critical cost drivers on one page, together with graphical and tabular outputs. The user can then try 'what if' scenarios by changing the values of the key cost drivers, receiving instant feedback of the effects on the financial results. A typical 'options' sheet is shown in Fig 8.2.

8.3   A Network for the 21st Century

As part of a programme to change BT's existing public network to new technology, it is envisaged that exchanges in the BT network could be updated to accommodate equipment that enables ATM and Ethernet connectivity in the local access network. Any of the approximately 5000 local exchanges could act as a serving exchange for this service, connecting access from the customer premises to BT's networks.

The cost effectiveness of providing access ATM and Ethernet connectivity has been investigated using cost-modelling and optimisation techniques. The basic idea is that deployment of new equipment to aggregate the ATM and Ethernet traffic at exchange locations will only cost in if enough traffic is generated at these nodes. To

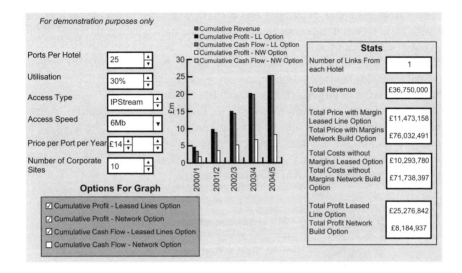

Fig 8.2   Options sheet for 'what-if' analysis.

generate sufficient traffic, it will be necessary to include the traffic from multiple customers, the cost benefit resulting from the costs of equipment being apportioned among the various customers. The modelling helped to identify the exchanges to be upgraded.

Mobile networks running second generation services, such as voice and text messaging, only require conventional equipment (TDM, PDH, SDH) to aggregate traffic. This capability can be realised using what is known as ASDH1 equipment. Third generation mobile services will require ATM connectivity (with ASDH2 equipment). Further generators of traffic include schools, which need access to the Internet.

In order to measure the cost effectiveness of supplying ATM and Ethernet access equipment (ASDH2) at the edge of the network, traffic levels from a number of types of customer were modelled — three mobile customers, along with secondary and primary schools. The geographical locations of cell sites and schools was important in this exercise as this provided accurate information about exactly which exchanges carried high levels of traffic and connected the most number of customers. Using this information, a list of exchanges that were potentially upgradable could be identified.

The exchanges were categorised into three types:

• those that already had ATM and Ethernet (ASDH2) equipment;

• those that had only conventional (ASDH1) equipment;

• those that had no ASDH equipment at all.

This categorisation of exchanges was important as it determined the relevant BT network costs required to connect a customer to the right equipment. It also defined the type of electronics required for the link between nodes.

Once the busy nodes were identified, a set of exchanges could be listed as potential upgrade exchanges. In essence, any exchange that handled above a certain threshold of traffic or customers was a candidate for an upgrade.

So, for example in the graph shown in Fig 8.3, we can identify 'isolated' exchanges (where there is no ASDH equipment available at these sites) and exchanges without ATM/Ethernet equipment, but where it could be provided, that carry a high level of traffic, e.g. greater than 35×2 Mbit/s links. These are potential candidates for an upgrade to ASDH2 exchanges.

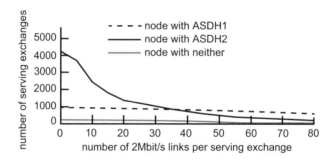

Fig 8.3   Choosing the largest exchanges for an upgrade.

The way these potential exchanges should be upgraded can be guided using various optimisation algorithms. One such approach would be to upgrade an exchange according to some survival-of-the-fittest principle. For example, the busiest exchange could be upgraded first and the corresponding network costed. Then the next busiest node could be costed, and the corresponding network costed again. Plotting the costs would then reveal which one comes in at least cost. This approach is fairly simplistic and would not necessarily lead to the minimum cost network. More complex algorithms can be used to home in on the minimum cost for this specific type of optimisation problem, providing a cost approaching the theoretical minimum. As the algorithm also gives the configuration of the network, the exchanges to be upgraded can be identified.

Using different scenarios for traffic generation by customers, it is possible to cost an optimised version of the network and even find configurations of the network that cost in lower than the option where no exchanges are upgraded. In this way, the roll-out of ATM/Ethernet access equipment can be guided in the most cost-efficient manner.

As telecommunications equipment evolves, operators need to upgrade their networks. However, as money is a limited resource in business, operators need to ensure, as best they can, that they minimise the costs of any equipment upgrade.

This can be done in two ways. Firstly, to upgrade equipment in certain locations so that the overall effect of upgrade is financially favourable. Secondly, to upgrade in a timely fashion so that equipment is bought as late as possible so that money can be working for the company for as long as possible, but not spent so late that it interferes with the network's grade of service. The following sections explore these twin strategies in more detail.

8.4   Optimisation

When a new technology comes along, it is unlikely that that it will be rolled out across all nodes in a network at once. It would not make financial sense. It is more prudent to roll out the equipment over a subset of nodes and phase the financing over time.

Figure 8.4 shows a network operator linking two customer sites via several nodes in its own network. The customer sees only a single link from site A to site B. The operator organises links so that nodes containing the relevant equipment for the link type are used. For example, the exchanges numbered 1, 2 and 3 reflect different technology roll-out at the site. All links go via a type 1 exchange, so the type 2 and 3 exchanges are routed to the type 1 exchange, before being routed through the other nodes in the operator's network, and then on to the customer site B.

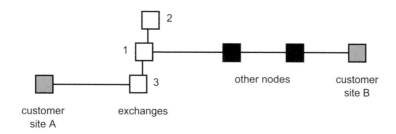

Fig 8.4   Customer connection routed via different technologies.

We can upgrade either of the exchanges of type 2 or 3 to a type 1. In doing this, we find that the route between sites A and B changes (Fig 8.5). The type 3 exchange is upgraded to a type 1, so the existing link to a type 1 is no longer required. Instead, the link to the next node in the network is shown as a dotted line. In fact, several parts of the network are affected in terms of routing and cost. Whether or not the overall effect of upgrading the exchange is good or bad depends on several factors, not least of which are traffic, routing, and location of nodes. All of these can be modelled and a figure for the cost of upgrading a particular exchange can be found.

The difficulty arises when networks contain several thousand nodes. Upgrading nodes can have a beneficial effect (if there is enough traffic through a node, an

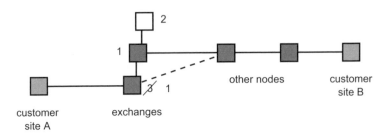

Fig 8.5   Customer connection after an exchange upgrade.

upgrade ends up with the network costing less), or a non-beneficial effect (an upgrade may increase the overall cost of a network). Combinations of upgrades complicate matters, as this may result in certain exchanges being upgraded as part of a group of others (as this would reduce costs), but would not be a candidate for upgrade in isolation, as this would increase cost. The number of different combinations of upgrades quickly becomes unmanageable, in even a small-scale network, as the computing power required to calculate all different options is well beyond the capabilities of even the most powerful computer we have today. For this reason, algorithms are employed to generate an answer, invariably not the ideal answer, but something within a few percent of the ideal.

Algorithms work by searching out an answer given a set of rules. Algorithms that have a random component associated with them give a useful first approximation to a solution.

For example, we can randomly select exchanges to upgrade, and keep the upgrade if costs reduce, and reject it if costs increase. The process can be compared to the diagram shown in Fig 8.6.

Imagine each point on the mesh represents a particular subset of exchanges that are upgraded. A cost is associated with each point. Some of the costs are lower (shown by the dips in the surface of the mesh). Now imagine randomly selecting a point on the mesh (representing a set of exchanges to upgrade), and then moving over the mesh until you find a neighbouring mesh point with a lower value. Repeat this until there are no neighbours with a lower value and you will have found the bottom of one of the holes (the dark arrows in the diagram show this process). It is like rolling a marble across an uneven surface and waiting for it to come to rest.

The algorithm stops when it has found the bottom of one of the holes. This relates to a minimum cost for the network and is arrived at by the algorithm selecting certain specific exchanges to upgrade. As there can be a random element to these algorithms, different runs can produce different answers resulting in different cost savings that can be made (see Fig 8.7).

These kinds of algorithm can guide new technology roll-out, identifying sites where new equipment should be deployed, while at the same time ensuring the lowest cost option.

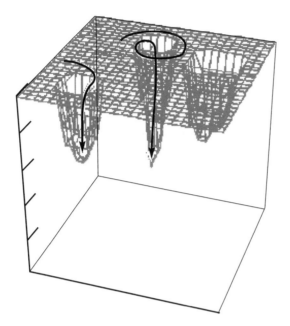

Fig 8.6   Optimisation algorithm finding cost minima.

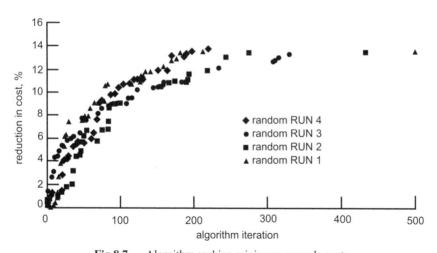

Fig 8.7   Algorithm seeking minimum upgrade costs.

These techniques can be used whenever there are many different options open to a network operator, for example when looking at upgrading nodes in a network as above, or when looking for aggregator sites within a network or when combining several networks. Such techniques are used to provide a framework and detailed guidance for network planners.

8.5   Risk Analysis

In conducting risk analysis, the uncertainties that companies or individuals face in their everyday business need to be investigated. When business performance is modelled, it is essential to look at various scenarios to identify unfavourable risk ones. Efficient risk management aims to identify all possible sources of risk. Having identified these sources, appropriate measures can be undertaken to reduce the risks involved. Inadequate risk quantification can lead to considerable losses, or even bankruptcy for the company.

A network operator can face a number of potential risks in their daily operations. For instance, the risk of traffic congestion or failure of equipment in their networks exposes them to the risk of offering poor service quality to their customers, with potentially far-reaching knock-on effects, when dissatisfied customers take their business elsewhere. To allay these risks, some network operators may install a large amount of capacity in their network. However, some of the capacity may remain idle for long, continuous periods of time. From an operational point of view, the operator has access to a commodity that is not being used. Thus, the operator would have made an investment in capacity which has not given them the maximum possible returns. Network operators face uncertainties regarding the future need for bandwidth to provide a high quality of service to their customers. The network operator wants to buy enough capacity to provide quality of service, but not buy so much that it remains idle. The network operator is being exposed to a certain risk which can be managed to maximise profits.

Many different approaches exist to forecast traffic. One novel method is a using Brownian motion formula to facilitate the prediction. This is one approach used in modelling work carried out by BT. Any expected decrease in the cost of equipment over time is forecast, as well as the increase in capacity. Historical data is used in both cases to predict future values. When the two predictions are combined, a graph of the type shown in Fig 8.8 can be generated.

This price-of-risk curve gives an indication of the savings that can be made by accepting more risk involved in delaying investment in capacity [2]. If we wait until we are 10% likely to run out of capacity, we could save 12% of our equipment cost. If we waited until we were 65% likely to run out of capacity, we could save 30% of our equipment costs.

8.6   Return on Investment in Network Vendor's Equipment

Apart from deciding when to invest in new equipment, network operators are also faced with decisions as to which vendor's equipment to buy and how to integrate it into existing operations. This is part and parcel of designing and evolving a telecommunications network, and contributes to the total cost of ownership of the network.

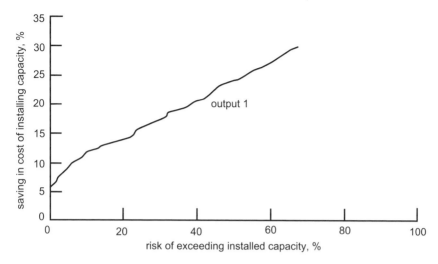

Fig 8.8   Savings made by accepting risks.

There are four key questions.

- How do capital purchase costs compare?

- What about the operational expense?

- Can any one vendor offer equipment that enables a step change in terms of the services that can be offered and markets that can be addressed, or the way the network is run?

- What will any whole-life cost savings be, and what return on the investment will the carrier see?

Looking at the other side of the coin, suppose a company, called Vendor Networks, is developing a new IP device that will support a comprehensive range of features. How can Vendor Networks develop a compelling value proposition that reveals the business benefits of the product to a network operator (NetCo)?

BT designs and runs TCO and ROI modelling activities to enable vendors to demonstrate the economic advantages of their product to customers who are network operators. All activities are carried out strictly under non-disclosure agreements. This ensures that the modelling results are only available to other parties at the request of the vendor.

A starting point is to develop an understanding of what the main capital (capex) and operational (opex) costs are for NetCo in running an advanced, feature-rich IP network (i.e. offering VPN and QoS capabilities). Only then is it possible for Vendor Networks to create the reasoning for a value proposition.

The crux of the issue is to investigate the implications of integrating the product into an existing network. Is it a simple matter of changing one type of kit for another, or are the consequences further-reaching than that? If Vendor Networks' product offers radically different capabilities or capacity, would it be possible to dramatically change the network structure, and strip out whole layers of equipment? What other activities will need to be undertaken by the carrier? What would it all cost? While these issues are not of direct interest to Vendor Networks itself, they are extremely important to its prospective customer. NetCo will inevitably be reluctant to change the *status quo*. The work involved in integrating new components into a network is rarely simple, and the fear of disrupting operations and adversely affecting service to NetCo's customers is always a barrier to the purchase of new hardware. Understanding this process can be key to making a sale.

BT has experience of identifying exactly how a vendor's product will affect the structure and operations of a telecommunications network, and how this translates into which network capital and operational cost elements are most affected. Highlighting the most significant changes gives the basis of the value proposition, and articulating it means being able not only to model 'before' and 'after' network scenarios, but also to quantify the changes in whole-life costs. This contrasts the size of the investment facing NetCo, with the all-important return on that investment.

Figure 8.9 shows a typical annual cost breakdown of the principal network capex costs. It is worth noting that while a slice of the pie may be a small percentage of total costs, this can nevertheless equate to many millions of pounds/euros/dollars. The 'before' and 'after' pie charts represent the relative total spend and different cost breakdown of network expenditure, following investment in new and more efficient technology.

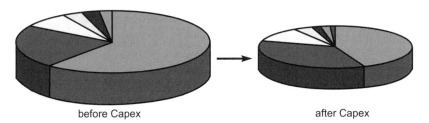

before Capex   after Capex

Fig 8.9   Typical changes in annual capital expenses in a network, following new
technology investments.

Similarly, Fig 8.10 represents the annual network spend on operational costs, and the changes that result from gains in operational efficiencies that become possible with investment in new-generation technology.

Vendor Networks may also wish to develop a ROI modelling tool, for use in discussions with NetCo and other prospective customers. Tools of this nature encapsulate the ROI value proposition, quantifying what the financial benefits and return on investment in the new product could be for that particular customer's circumstances. The tool invites the customer, NetCo, to specify values of key cost drivers, and to see for itself, via a simple calculator, what cost savings can be achieved (see Fig 8.11).

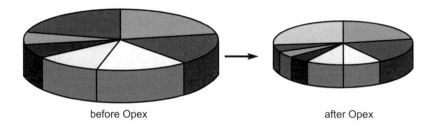

before Opex   after Opex

Fig 8.10   Typical changes in annual network operational expenses, enabled by new
generation technologies.

In order to understand and articulate to prospective customers the benefits of
purchasing Vendor Networks' product in particular, BT is also able to carry out
comparisons of return on investment with other vendors' products. Obviously, such
calculations can only be carried out on publicly available data of products from
Vendor Networks' rivals, and can provide a useful supplement to Vendor Networks'
value proposition.

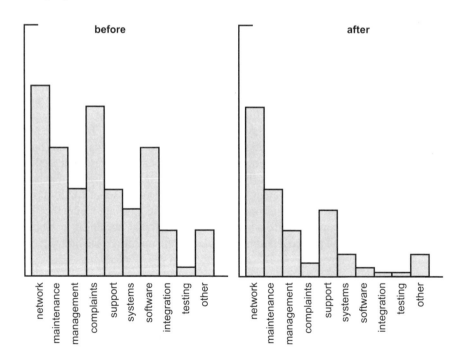

Fig 8.11   Achievable total cost of ownership savings.

8.7   Mobility ROI Model — Openzone Wireless LAN

ROI modelling can be about cost savings, or it can be about investment to generate new revenue streams from new products and services. This section outlines some ROI modelling work in support of a full business case for BT's UK-wide public wireless LAN access service, Openzone.

Following the demerger of O_2 from the BT Group, BT was looking for opportunities to strengthen its presence in the mobility market. Here mobility means much more than just providing wireless access to handsets; it is about providing services to customers on the move, wherever they are, using whatever technology is best suited to the task and is most cost effective.

BT Openzone is one mobility proposition that provides access to intranets and the Internet for people with laptops and PDAs, etc, who are on the move. It is available at locations where business users who are travelling pause for a while, and sit and wait, e.g. airports, cafes, and railway stations. It offers 500 kbit/s access based on IEEE802.11b (2.4 GHz).

BT was asked to assess the business case and financial viability of the service before it was launched. The emphasis was on providing consultancy based on a tool which covered a comprehensive range of business and technology issues, which enabled extensive sensitivity studies, and which was to provide input to the business case as a whole.

Working with both infrastructure designers and marketing people an extremely robust and flexible model was developed, suitable for investigating a range of 'what if' type scenarios. Such scenarios considered the number and size of sites, site roll-out strategies, the mix of transmission technologies, and cost changes over time. This particular work did not cover financing deals or risk sharing and indemnity payments between the various players in the business, but it did include investigating pricing ideas and potential revenues. The model calculated the capital expenditure, operational (current account) costs, and revenues. Investment analysis techniques were applied to the cash flows to ascertain ROI measures — payback, internal rate of return , and net present value (NPV) — for each set of conditions.

Needless to say, the financial conclusions were just one input into the final design of the service and its tariff scheme — organisational, technical feasibility, and marketing issues all played a part in formulating the initial service offering, and in its appeal to customers.

8.8   Summary

This chapter has presented a brief overview of some of BT's network cost optimisation and return on investment modelling capability. It has outlined a number of modelling techniques, and shown how they have been applied to evaluate the potential of a range of new opportunities for developing BT's networks and services. It also shows how BT has been working with vendors to evaluate and communicate the promise of next-generation network equipment.

References

1   Operis TRG Limited, http://www.operis.com

2   Olafsson, S.: '*Managing Risk in Network Capacity Investments*', Complexity Research Group Paper, BT.

9

A NEW APPROACH IN ADMISSION CONTROL AND RADIO RESOURCE MANAGEMENT FOR MULTISERVICE UMTS

P Shekhar, S Abraham, A Rai and S Devadhar

9.1   Introduction

The Universal Mobile Telecommunications System (UMTS) has been designed to support real-time services including both multimedia and packet data services. Multimedia in UMTS means that the simultaneous transfer of speech, data, text, pictures, audio and video with a maximum data rate of 2 Mbit/s will be possible. Simultaneous use of several applications raises the demands for mechanisms which can guarantee quality of service (QoS) for each application. A UMTS bearer-service layered architecture is shown in Fig 9.1; each bearer service on a specific layer offers its individual services by using the services provided by the layers below [1].

UMTS provides several radio resource management (RRM) [2] strategies to the QoS requirements. Some services do not have stringent delay requirements. This opens up the area of using the radio resources efficiently while guaranteeing a certain target QoS and maintaining the planned coverage and capacity of the network. Within the UMTS bearer service, the radio bearer service covers all aspects of radio interface transport [3]. The main focus of this chapter is the RRM strategies used in the radio bearer service.

The radio interface architecture is layered into three protocol levels:

- the physical layer (L1);
- the data link layer (L2);
- the network layer (L3).

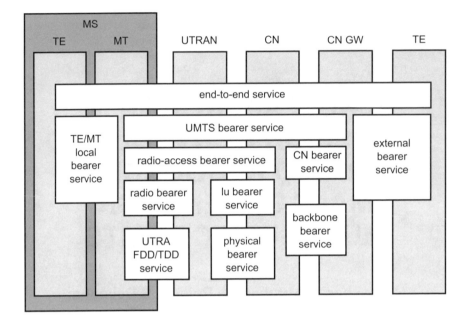

Fig 9.1   UMTS QoS architecture.

Radio resource control (L3) controls the configuration of L1 and L2. One of the important functions of the MAC [4] layer is to select an appropriate transport format for each transport channel depending on the instantaneous source rate (see Fig 9.2).

This chapter presents an RRM algorithm which tries to increase the capacity of the system by reducing instantaneous interference through the use of lower data rates, while still maintaining the QoS requirement of the interactive and background services for an individual mobile user in a UMTS network. The scope of this work is based on the following three premises:

- radio resource management strategies are not subject to standardisation;

- the air interface transmission can be dynamically optimised only by changing the transport formats in the transport format set, without any problems of resetting RLC, losing data, or retransmitting [5];

- as per 3GPP TS 25.321 V5.4.0 (2003-04), the details of the computation of the available bit rate in the MAC layer and its interaction with the application layer are not specified.

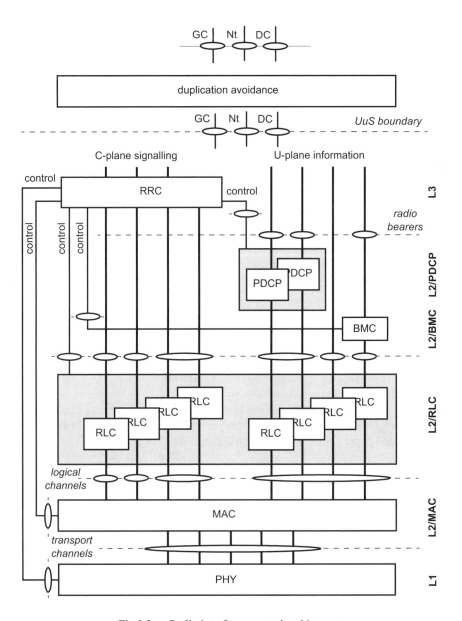

Fig 9.2 Radio interface protocol architecture.

9.2   Background

In this section the state-of-the-art algorithms are explained in detail. Section 9.2.1 explains the service credit (SCr) algorithm, with section 9.2.2 describing the MR algorithm, and section 9.2.3 providing a detailed review of these algorithms.

9.2.1   Service Credit Algorithm

This algorithm [6] aims to offer a negotiated average bit rate. It accounts for the difference between obtained bit rate and guaranteed bit rate. If the obtained bit rate is higher than the guaranteed bit rate then service credit is decreased to reduce the data rate of the connection. If the obtained bit rate is less than the guaranteed bit rate, then service credit is increased to enhance the data rate.

9.2.1.1   *Input*

The following data is used in the process:

TB_{size} — number of information bits sent in a transmitted block;
TB_{max} — number of transport block corresponding to the transport format TF_{MAX};
TTI duration — duration of a transmission time arrival;
L_b — number of bits waiting for transmission in buffer;
Guaranteed_rate — it accounts for the bit rate to be guaranteed to the user;
Transmitted_TB (n–1) — number of transport blocks that have been successfully transmitted in the previous transmit time interval (TTI).

9.2.1.2   *Process*

There are three process steps for the *n*th TTI.

Step 1:
/*Compute the service credit for every TTI. */
If $(n == 0)$
{
  $SCr(n) = 0$
}
else
{
  $SCr(n) = SCr(n-1) + Guaranteed\_rate/TB_{size} - Transmitted\_TB(n-1)$
}
                    (...... 9.1)

Step 2:
/* Compute the number of transport blocks to be transmitted.*/
$$numTB = min\ (L_b/TB_{size},\ SCr(n),\ TB_{max}) \qquad\qquad (......\ 9.2)$$

Step 3:
Set the TF format according to the number of transmitted block (*numTB*)
calculated as shown in the following steps.
If (*numTB* > 0)
{

       *TF = minimum transport format within a TFS that allows to transmit
       numTB transport blocks.*

}
else
{

       TF = 0

}   (...... 9.3)

9.2.2   Maximum Rate (MR) Algorithm

This algorithm [6] selects the TF format that allows the highest transmission bit rate
according to the number of bits in the buffer waiting for transmission. It checks the
buffer occupancy in every TTI and sends the data at the higher transmission bit rate.
If the buffer occupancy is less, it selects the lower TF formats, otherwise it always
sends the data at the highest transmission rate.

9.2.2.1   *Input*

The following data is used in this process:

TB_{size} — number of information bits transmitted in a transmitted block;
TB_{max} — number of transport block corresponding to the transport format TF_{MAX};
TTIduration — duration of a transmission time arrival;
L_b — number of bits waiting for transmission in buffer.

9.2.2.2   *Process*

In the *n*th TTI, there are two process steps.

Step 1:
/* Compute the number of transport blocks to be transmitted */
$$numTB = min\ (\ (L_b/TB_{size}),\ TB_{max})) \qquad\qquad (......\ 9.4)$$

Step 2:
Set the TF format according to the *numTB* value calculated in Step 1.
If (*numTB* > 0)
{

       TF = minimum transport format within a TFS that allows to transmit
       numTB transport blocks

}
else
{

       TF = 0

}   (...... 9.5)

9.2.3   Review of Algorithms

The RRM algorithms employed in the user equipment–MAC layer are evaluated by their transport format usage. UE-MAC decides the selection of the TF (transport format) from a given TFS (transport format set). Analysis [6] of the results shown in Fig 9.3 shows that SCr24 (SCr24 means guaranteeing a rate of 24 kbit/s) uses TF1 and TF2 more than TF3 and TF4 by guaranteeing a transmission rate of 24 kbit/s. The SCr algorithm gains credit when the UE buffer is empty, and if a new packet arrives it uses a higher TF format (TF3 and TF4), increasing the transmission rate over the guaranteed one.

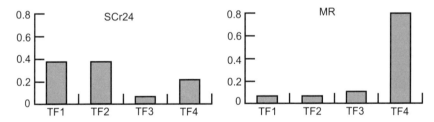

Fig 9.3   Comparison of TF distributions for the SCr24 and MR RRM algorithms.

Referring to Fig 9.3, the MR algorithm tries to transmit the packets as fast as possible and thus it tends to use TF4 most of the time. MR takes the buffer occupancy into account. By maximising the transmission rate, the MR algorithm increases the interference caused by a mobile to the system. On the contrary, since SCr does not take buffer occupancy into account, it provides better control of the transmission rate.

SCr guarantees a certain mean bit rate. Our proposed algorithm takes buffer occupancy into account and tries to use a lower transmission rate while meeting the QoS requirements of the interactive and background services. By lowering the transmission rate and trying to use a middle path between MR and SCr, our algorithm tries to maintain balance between capacity and quality of system. The re-lation between interference and transmission rate is explained in section 9.2.4. As the algorithm sits in UE RRM, we are only addressing the issue of uplink interference.

9.2.4   Relation between Capacity and Transmission Rate

The uplink signal to interference ratio (SIR) for a mobile is the ratio between the signal strength of the received power and the sum of all signals received at the base-station (BS).

The equation for SIR is:

$$S/I = (E_b/N_o) \times (R/W) \tag{...... 9.6}$$

where:

S is the signal strength;

I is the strength of interfering signal;

E_b is the energy per bit of the signal;

N_o is the signal spectrum density;

R is the user data rate;

W is the UMTS-specific chip rate (3.8 MHz).

From equation (9.6) it is seen that S/I is directly proportional to R, the user data rate, which means that, if the transmission data rate is reduced, then the S/I produced by a particular mobile to the base-station will reduce. In UMTS the capacity of a cell is determined by system noise, thermal noise and interference. If a cell is serving N users, then the signal received at the BS for individual users means that it contains one desired signal of power S and ($N-1$) interfering signals with the identical power S (assuming ideal power control). In the simplest form, capacity is estimated as:

$$C = [W/(R.\alpha]/[E_b/N_o] \tag{...... 9.7}$$

where α is the voice activity factor.

From equation (9.7) it is clear that capacity is inversely proportional to data rate. It means capacity is soft and depends on multiple variables. This capacity is the capacity at the physical layer. The physical layer receives a set of transport blocks (TBs) from higher layers and the corresponding transport format indicator (TFI); these transport blocks are offered in predefined transport format, namely, TF0 to TF4 by L1 (layer 1) to MAC (media access control) and vice versa. This transport format consists of one dynamic part (transport block size, transport block set size) and one semi-static part (TTI). Based on the selection of transport format size, the data rate on the physical channel will vary. Table 9.1 contains the indicative figures used at the time of simulation. These figures are in line with the UMTS recommendation.

Table 9.1   Transport formats for the considered radio access bearers.

		Interactive class	Background class
TB size (bits)		300 bits	640 bits
TTI (transmit time interval)		10 msec	10 msec
TFS (transport format set)	TF0	0×300 (0 kbit/s)	0×640 (0 kbit/s)
	TF1	1×300 (30 kbit/s)	1×640 (64 kbit/s)
	TF2	4×300 (120 kbit/s)	10×640 (640 kbit/s)
	TF3	8×300 (240 kbit/s)	20×640 (1280 kbit/s)
	TF4	10×300 (300 kbit/s)	30×640 (1920 kbit/s)

9.3   Proposed Algorithm

We propose the Min-Max rate algorithm for interactive and background classes as it addresses the various issues related to both the MR and the SCr algorithms. This algorithm is used for radio resource management on the uplink channel for W-CDMA.

9.3.1   Min-Max Rate Algorithm

The algorithm determines the transport format (TF) in a TTI by taking into account the input traffic rate, buffer size and maximum number of transport blocks.

9.3.1.1   Input

The following data is used for this process:

TB_{max} — maximum number of TBs that the UE can transmit per TTI,

nT_s — indicates the nth TTI,

$E[nT_s, C]$ is the number of TB for the nth TTI (nT_s) and it is a function of traffic class denoted by C,

$N_b[nT_s, C]$ is number of bits available for transmission for nth TTI for traffic class C,

TB_{size} is number of bits per transmitted block (TB),

Current_change is the change in the buffer size in the current TTI,

$$(Current\_change = N_b[nT_s, C] - N_b([(n-1)T_s, C])$$

Previous_change is the change in the buffer size in the previous TTI,

$(Previous\_change = N_b([(n-1)T_s, C]) - N_b([(n-2)T_s, C]$

increment_step — increment value in the Δ_{TB},

decrement_step — decrement value in the Δ_{TB}.

9.3.1.2   Process

The algorithm provides the staircase approximation of the input information. The difference between input and output is quantified into multiples of $\pm\Delta$, which corresponds to positive and negative differences respectively. If the previous approximation lies below the current input information, then approximation is increased by certain multiples of Δ; but if, on the other hand, previous approximation lies above the current input information, it is decreased by certain multiples of Δ.

The algorithm calculates the number of transport blocks (TBs) to be sent in one TTI by accounting input-traffic rate, buffer size and maximum number of transport blocks. TB_{max} and the number of TBs available for transmission ($Nb[n]/TB_{size}$) limits the number of TBs per TTI.

Step 1:
Current_change = $N_b[nT_s, C] > N_b[(n-1)T_s, C]$ (...... 9.8)

Previous_change = $N_b[(n-1)T_s, C] > N_b[(n-2)T_s, C]$ (...... 9.9)

If (Current_change > Previous_change)
{
  $\Delta_{TB} = \Delta_{TB} + increment\_step$
}
else if (Current_change < Previous_change)
{
  $\Delta_{TB} = \Delta_{TB} + decrement\_step$
}

Step 2:
If (n == 0)
{
  $E[nT_s, C] = 0$
}
else if (n > 0)
{
  $E[nT_s, C] = E[(n-1)T_s, C] + \Delta_{TB}$

$$If\ (E[nT_s, C] > TB_{max})$$
$$\{$$
$$\qquad E[nT_s, C] = TB_{max}$$
$$\}$$
$$else\ if\ (E[nT_s, C] < 0)$$
$$\{$$
$$\qquad E[nT_s, C] = 0$$
$$\}$$
$$\}$$
(...... 9.10)

Step 3:
Calculate the number of TB to be transmitted.

$$num\_TBs(nT_s, C) = Minimum\ (TB_{max}, N_b[nT_s, C]/TB_{size}, E[nT_s, C])$$ (...... 9.11)

Step 4:
Set the TF format.

$$If\ (\ num\_TBs(nT_s, C) > 0)$$
$$\{$$
TF format = minimum transport format in the TFS that allows num_TBs(nT_s, C)
transport blocks.
$$\}$$
else
$$\{$$
$$\qquad TF = 0$$
$$\}$$
(...... 9.12)

9.4   Analysis Methodology

We have used a simulation approach to validate our algorithm and to compare it with the MR and SCr algorithms. The model consists of a traffic generator and a queuing model as shown in Fig 9.4.

We have used the OPNET tool to implement the different algorithms. The OPNET environment incorporates tools for all phases of a study, including model design, simulation data collection and data analysis. An OPNET model has three levels of hierarchy:

- network model level (a);

- node model level (b);

- process model level (c, d).

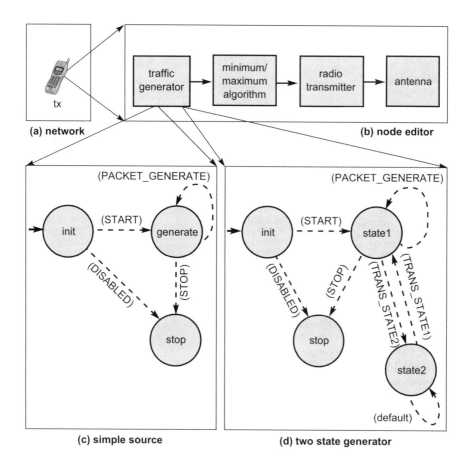

Fig 9.4 Network, node and process model for traffic sources (simple source, ON/OFF traffic generator).

9.4.1 Traffic Source Model

It has been widely demonstrated [7-9], that real-world traffic does not exhibit the ideal characteristics of the Poisson and Markov distributions previously used. As Sallent et al [6] do not provide the source models to generate the results of Fig 9.3, we have used the traffic models of section 9.4.1.2 for recreating that observation. Using this source traffic model we compare three algorithms.

The traffic sources used are:

• simple source traffic model with Pareto inter-arrival time;

• two-state traffic generator model with Pareto inter-arrival time in one state and constant inter-arrival time in the other state.

For the long-range dependent (LRD) traffic we use the Pareto distribution, since it has been shown to fit a number of traffic sources [9]. The Pareto distribution is defined by shape ($c > 0$) and location ($a > 0$) parameters, as defined in the following equation:

$$f(x) = ca^c/x^{c+1} \qquad\qquad a < x < \infty \qquad\qquad (\ldots\ldots 9.13)$$

Figure 9.4 shows the simple source traffic model (c) and two-state traffic generator process model (d).

9.4.1.1   Simple Traffic Generator

The simple traffic generator model has the following parameters:

- *inter-arrival time* — inter-arrival time of packets;
- *packet minimum size* — minimum size in bits of the packets;
- *packet maximum size* — maximum size in bits of the packets.

We have used Pareto distribution for inter-arrival time, and packet size is generated by uniform distribution with *minimum* as packet minimum size and *maximum* as packet maximum size. The process model for the traffic generator is also shown in Fig 9.4.

State Transition Description

The state transition conditions are given below.

- *START*

  This condition checks the arrival of the first packet. As soon as the event for the first packet is generated, control goes to *generate* state, which generates the packet and forwards it to the connecting module.

- *PACKET_GENERATE*

  This condition keeps track of packet inter-arrivals. After each inter-arrival time this condition becomes true and a new packet is generated and forwarded to the connecting module.

- *STOP*

  This condition holds true at the *stop time*. At the *stop time* packet generation stops and the control goes to *stop* state.

- *DISABLED*

    This condition is true if *stop time* is less than *start time*. Since this is a wrong entry by the user, control goes to the *stop* state from *init* state and no traffic is generated by the source.

State Description

The state descriptions are given below.

- *init*

    This state is the initialisation state, which is called once at the start of the simulation. This state intialises the *start time, stop time* and *packet size* parameters.

- *generate*

    This state generates the packet at every inter-arrival time.

- *stop*

    This state is the disabled state. No packet is generated in this state. Control comes to this state because of wrong entry or after the *stop time* has expired.

Model Description

init is the initialisation state. This state intialises the *start* and *stop times* and the *packet size* parameters. Whenever the event for the generation of the first packet is triggered, control goes to *generate* state, which is in charge of generating packets. After expiry of the *stop time*, control goes to *stop* state, where the system does nothing and the traffic generator goes into the *idle* state.

9.4.1.2 Two State Generator Model

This model requires the following parameters:

- *state1 duration (sec)* — time interval when the system remains in *state1*;
- *state2 duration (sec)* — time interval when the system remains in *state2*;
- *state2 inter-arrival time (sec)* — inter-arrival time of generation of packets in *state2*;
- *state1 inter-arrival time (sec)* — inter-arrival time of generation of packets in *state1*;
- *state2 packet max size (bits)* — maximum size of packet generated in *state2*;

- *state2 packet min size (bits)* — minimum size of packet generated in *state2*;
- *state1 packet max size (bits)* — maximum size of packet generated in *state1*;
- *state1 packet min size (bits)* — minimum size of packet generated in *state1*.

We use the Pareto distribution in *state1* for the inter-arrival times and constant distribution in *state2* for inter-arrival times. By using this generator, high-intensity traffic is generated when the control is in *state1* and low-intensity traffic is generated when the control is in *state2*. Results for different Pareto distributions have been listed in section 9.5.2.

State Transition Description

Descriptions for state transition for *start, stop, disabled* and *packet generate* are provided under the State Transition Description heading above in section 9.4.1.1.

- *TRANS_STATE1*

    This condition checks the times for transition from *state2* to *state1*. It ensures that the control goes to *state1* from *state2* after *state2* duration has expired.

- *TRANS_STATE2*

    This condition checks the times for transition from *state1* to *state2*. It ensures that the control goes to *state2* from *state1* after the *state1* interval has expired.

State Description

State descriptions for *init* and *stop* states are provided under the State Description section above in section 9.4.1.1. Functionality for *state1* description is similar to *generate*, which is described under the same heading.

- *state2*

    This state generates packets during *state2* duration at every *state2* inter-arrival time.

Model Description

init is the initialisation state. This state intialises the start and stop timing and the packet size parameters. Whenever the interrupt for the generation of first packet comes control goes to *state1* state, which is in charge of generating packets. After the expiry of *state1*, duration control goes to *state2*. *state2* generates packets at every *state2* inter-arrival time. After the expiry of *state2*, control goes to *state1*. In the similar fashion the system oscillates between *state2* and *state1*. After the *stop time* the control goes to *stop* state which does nothing and the system goes to *idle* state.

9.4.2   Queue Model

This finite state model (FSM) for the queue model is shown in Fig 9.5.

RRM algorithms (MR, Min-Max, SCr) have been implemented in *svc_queue* state, which is triggered at every TTI. Section 9.4.2.1 explains the transition conditions, section 9.4.2.2 explains the state description, and section 9.4.2.3 explains the FSM in detail.

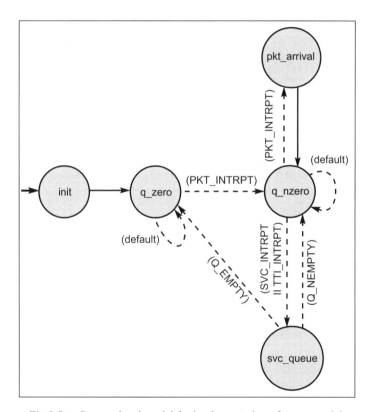

Fig 9.5   Process level model for implementation of queue model.

9.4.2.1   State Transition Description

The state transitions are described below.

- *PKT_INTRPT*

  This condition checks the arrival of the new packet.

- *SVC_INTRPT*

  The FSM model uses the lazy scheduling. In this process the system does not generate TTI interrupt when buffers of all the traffic classes are empty. During this time the system waits in *q_nzero* state for the arrival of packet. Whenever a new packet comes an event is generated at the next *TTI*. *SVC_INTRPT* condition checks for this event to send the control to *svc_queue* state.

- *TTI_INTRPT*

  This condition checks the TTI interrupt which is generated by *svc_queue* state after every TTI time.

- *Q_EMPTY*

  This condition checks the data in the buffer. If there is no data available in any queue, then *Q_EMPTY* condition holds true. This condition is used for the transition from *svc_queue* state to *q_zero* state.

- *Q_NEMPTY*

  This condition checks buffer data available in the queue in all the traffic classes. If there is data available in any queue, then *Q_NEMPTY* condition holds true. This condition is used for the transition from *svc_queue* state to *q_nzero* state.

9.4.2.2   State Description

The state descriptions are given below.

- *init*

  This is an initialisation state, which is called once at the start of the simulation. This state intialises the queue, statistics and algorithm-related variables.

- *q_zero*

  This is the state when the buffer is empty and there is no data to transmit. This state waits for the input traffic.

- *q_nzero*

  This is the state when the buffer has data to transmit. This state waits for the arrival of a new packet and the event for the next TTI. If the new packet comes, control goes to *pkt_arrival* and, if a TTI event comes, control goes to *svc_queue* state.

- *pkt_arrival*

  Whenever a new packet arrives, control goes to *pkt_arrival* state. Different queues are maintained for different traffic classes. This state checks the traffic class of the input traffic and inserts the packet in the appropriate queue, depending on the class. If the control has come from *q_zero* state, then this state generates the service interrupt for the next nearest TTI.

- *svc_queue*

  Control comes to this state when either *SVC_INTRPT* or *TTI_INTRPT* condition is true. This state does the following:

  — calculates the number of transmitted blocks for the current TTI based on the algorithm used (Min-Max rate, MR, SCr algorithm);

  — selects the TF from the transport format set (TFS) based on the number of the transport block calculated.

  In the end this state checks the buffer size and if the buffer is empty, then control goes to *q_nzero* state, otherwise control goes back to *q_zero* state.

9.4.2.3   Model Description

q_zero is the state when the system has no data to transmit. Whenever a packet comes, it is inserted into the buffer and the control goes to *q_nzero* state. *q_nzero* waits for the new packet and the event to indicate the start of a new TTI. When a new packet comes, control goes to *pkt_arrival* state. *pkt_arrival* state inserts the packet into the buffer. When the event for the TTI comes, the control goes to *svc_queue* state. *svc_queue* state is responsible for the calculation of the number of TBs based on the algorithm used (MR, SCr and Min-Max) and their TF mapping. If the buffer is empty, control goes to *q_nzero* state, otherwise control goes to *q_zero* state. In the *q_zero* state the system waits for the arrival of a new packet and the event to indicate the start of a TTI.

9.5   Results and Analysis

Section 9.5.1 explains the simulation model validation and in section 9.5.2 the simulation analysis is explained. Section 9.5.2 also shows the comparison of the TF usage pattern of the MR, SCr and Min-Max algorithms for different input traffic distributions.

Section 9.5.3 provides analysis of these plots and compares the Min-Max algorithm with the MR and SCr algorithms.

9.5.1   Simulation Model Validation

The simulation model validation is done using the Naylor and Finger approach:

- the model has been developed with high face validity;

- validation of the model has been done using various structural assumptions, with the model doing the task of TF selection strictly with the following assumption:

    — the air interface transmission can be dynamically optimised only by changing the transport formats in the transport format set, without any problems of resetting RLC, losing data, or retransmitting [5];

    — as per 3GPP TS 25.321 V5.4.0(2003-04), the details of the computation of the available bit rate in the MAC layer and its interaction with the application layer are not further specified;

- the model input-output is compared with the expected behaviour of the model, which is to take the input traffic rate and buffer occupancy into account and use TF2 and TF3 formats most of the time.

9.5.2   Simulation Results

In this section the results of interactive class for the MR, Min-Max and SCr algorithms are shown. Then the results of background class are shown, which are analysed in section 9.5.3. Figures 9.6-9.9 show the TF distribution of the SCr, MR and Min-Max rate algorithm for the interactive and background class (e-mail application) with different distributions of the packet inter-arrival time by different traffic source models.

Figures 9.10 and 9.11 show the average packet end-to-end delay for the MR and Min-Max (Fig 9.10) and SCr (Fig 9.11) algorithms for interactive class for Pareto distribution (0.005,5). It is clear from these that delay for Min-Max is higher than MR but lower than SCr. In Figs 9.10 and 9.11, the X-axis shows the simulation time (min) and the Y-axis shows the average end-to-end delay (sec).

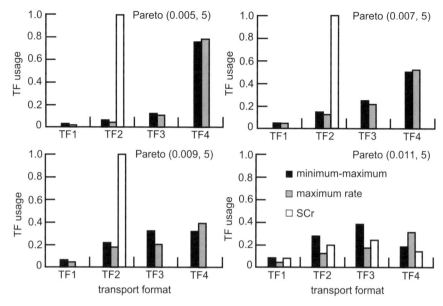

Fig 9.6   TF usage comparison for the MR, Min-Max, SCr algorithms for interactive class by using the simple source generator.

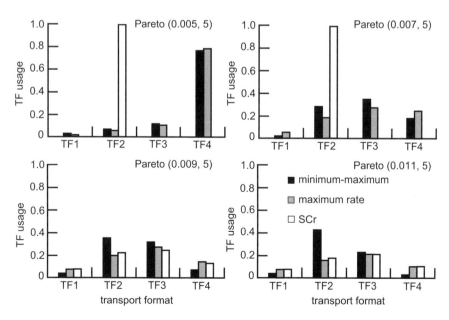

Fig 9.7   TF usage comparison for the MR, Min-Max, SCr algorithms for interactive class by using the two state generator.

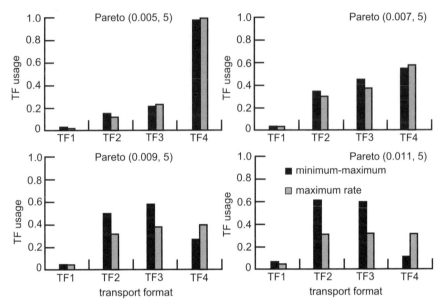

Fig 9.8   TF usage comparison for the MR and Min-Max algorithms for background class
by using the simple source generator.

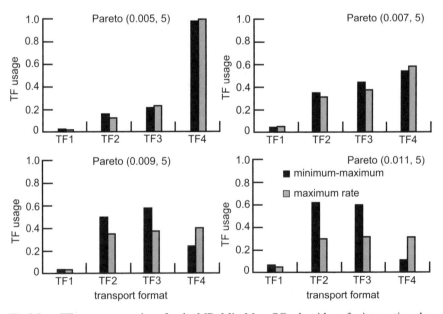

Fig 9.9   TF usage comparison for the MR, Min-Max, SCr algorithms for interactive class
by using the two state generator.

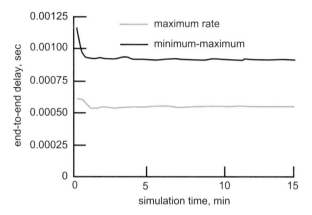

Fig 9.10   Average packet end-to-end delay for interactive class for Min-Max and MR for Pareto (0.005, 5) distribution.

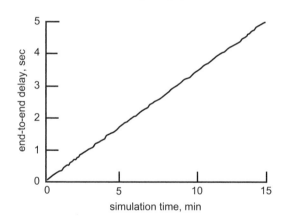

Fig 9.11   Average packet end-to-end delay for interactive class for SCr for Pareto (0.005, 5) distribution.

9.5.3   Analysis

This section explains the results of the MR, Min-Max and SCr algorithms shown in section 9.5.2 and then provides a comparison of the results of the Min-Max algorithm with those of the MR and SCr algorithms.

9.5.3.1   MR Algorithm

The MR algorithm has been described in detail in section 9.2.2. This algorithm takes buffer occupancy into account and sends the data at the highest transmission rate possible. So in MR, TF4 format usage is higher than with other formats when buffer occupancy is high.

As the buffer occupancy decreases, lower TF formats are selected and this explains the selection of lower TF formats in the MR algorithm. Since buffer occupancy is directly proportional to input traffic, it can be concluded that TF4 usage should be high for high input traffic and TF4 usage should decrease as the input traffic rate decreases.

Results for the MR algorithm are shown in Figs 9.6-9.9, where the LRD source has high-intensity traffic (Pareto (0.005, 5)) and where TF4 usage is high in the MR algorithm. As the input traffic decreases, TF4 usage will decrease and usage of other lower TFs will increase.

9.5.3.2   SCr Algorithm

Details of the SCr algorithm have been given in section 9.2.1. This algorithm transmits the data at the guaranteed rate. When the transmitted rate is greater than the guaranteed rate, service credit is decreased to reduce the transmission rate and in similar fashion if the transmitted rate is less than the guaranteed rate, service credit is increased to enhance the transmission rate. So the transmission rate always remains near the guaranteed rate and this explains the usage of TF1 and TF2 formats most of the time. During the period when the buffer is empty (idle period), service credit increases continuously as there is no upper limit for service credit. This causes the increase in the transmission rate (above the guaranteed rate) when a new packet arrives after the idle period and this explains the usage of higher TFs (TF3 and TF4).

The results of SCr are shown in Figs 9.6 and 9.7 from which it is clear that TF2 usage is high for the higher input traffic. When the input traffic rate reduces, the TF2 usage will decrease and it will spread to TF3 and TF4 which can be seen for Pareto (0.011, 5) distribution in Fig 9.6 and Pareto (0.009, 5) and Pareto (0.011, 5) in Fig 9.7.

9.5.3.3   Min-Max Algorithm

The Min-Max algorithm is explained in section 9.3.1. This algorithm takes input traffic rate and buffer occupancy into account and reduces the transmission below the highest transmission rate. It uses the staircase approximation to gradually increase or decrease the transmission rate based on the input traffic rate. If the input traffic increases, it does not jump to the highest transmission rate as does the MR algorithm, but rather increases the transmission rate in steps and thus the growth is gradual. Due to this, TF4 usage in the Min-Max algorithm is less than in the MR algorithm and its TF usage lies mostly in TF2 and TF3 format. This result can be seen for the Min-Max algorithm for Pareto distribution (0.007,5), (0,009,5) and (0.011,5) in Figs 9.6-9.9. These results show that TF2 and TF3 usage is higher in the Min-Max algorithm. As the input traffic rate increases, the TF4 usage will increase but it will always be less than the TF4 usage of the MR algorithm.

This result can be seen in Figs 9.6-9.9 for Pareto distribution (0.005,5) where TF4 usage is higher than TF2 and TF3 transport format.

Comparison with MR Algorithm

Results of the Min-Max algorithm are shown in Figs 9.6-9.9. They show that in the Min-Max algorithm, TF4 usage is high when the input traffic is high (Pareto (0.005, 5)), but TF4 usage of the Min-Max is less than the MR algorithm. On the other hand, TF2 and TF3 usage is higher in the Min-Max algorithm when compared to the MR algorithm. The same trend is followed for a reduced input traffic rate. Here TF2 and TF3 usage for the Min-Max is higher than for the MR, but the TF4 usage is always lower than the MR algorithm. Since the Min-Max algorithm uses lower TF formats, the data rate for the Min-Max algorithm is less than the MR algorithm and hence the delay of the Min-Max algorithm is greater than the MR algorithm. This explains the delay comparison graph (Fig 9.10) for the Min-Max and MR algorithms where delay for the MR is less than the Min-Max algorithm.

Comparison with the SCr Algorithm

As expected, TF2 usage for the SCr algorithm is always greater than for the Min-Max algorithm. This result can be seen for Pareto distribution (0.005,5), (0.007,5), and (0.009,5). As the input traffic rate reduces, the idle period increases; TF2 usage decreases and TF usage is spread across all TF formats. This explains the data in Fig 9.6 for Pareto distribution (0.011, 5) and Fig 9.7 for Pareto distribution (0.009,5) and (0.011, 5) where SCr is spread across all TF formats. The SCr algorithm does not check the buffer occupancy and sends data at a guaranteed rate. So the data rate of the SCr algorithm is less than for the Min-Max algorithm. This explains the lower delay for Min-Max compared to SCr algorithm. Since SCr does not respond to the buffer occupancy but maintains a relatively stable rate, it can cause large delay in the system when the buffer occupancy is high; this result is shown in Fig 9.11, where it is shown that the delay for the SCr algorithm increases continuously.

9.6   Summary

In conclusion, it can be seen from the simulation results that the SCr algorithm keeps gaining credit when the buffer is continuously empty. The SCr algorithm has a limitation in that it does not define the upper limit of the service credit. The SCr behaviour (as stated in Sallent et al [6]) of using TF1 and TF2 when the buffer occupancy is high, and the usage of TF3 and TF4 as a new packet arrives when a buffer is empty, is exhibited only for a particular type of traffic source. MR uses the highest possible transmission rate most of the time and thus increases the interference caused by a mobile to the system. As a consequence, MR tends to decrease the system capacity.

The Min-Max rate algorithm uses the available middle order TFs. It uses TF2 and TF3 format most of the time, for background and interactive class of service, as compared to the MR and SCr algorithms. So the data rate of the Min-Max algorithm is lower than that for an MR algorithm. As a lower data rate increases the capacity of the system (see section 9.2.4), it can be concluded that system capacity will be higher by using the Min-Max algorithm compared to the MR algorithm.

References

1   3GPP TS 23.107 V5.8.0: '*Quality of Service (QoS) Concept and Architecture*', — http://www.3gpp.org/

2   3GPP TR 25.922 V4.0.0: '*Radio Resource Management Strategies*', — http://www.3gpp.org/

3   3GPP TS 25.301 V5.2.0: '*Radio Interface Protocol Architecture*', — http://www.3gpp.org/

4   3GPP TS 25.321 V5.4.0: '*Medium Access Control (MAC) Protocol Specification*', — http://www.3gpp.org/

5   3GPP TSG RAN WG2#1 Tdoc RAN WG2035/99, — http://www.3gpp.org/

6   Sallent, O., Perez-Romero, J., Agusti, R. and Casadevall, F.: '*Provisioning Multimedia Wireless Networks for Better QoS: RRM Strategies for 3G W-CDMA*', IEEE Communications Magazine (February 2003).

7   Jung, S. Y., Hong, J. H. and Kim T-S.: '*A Formal Model for User Preference*', IEEE Conference on Data Mining (2002).

8   Kozlovski, E., Duser, M., Killey, R. I. and Bayvel, P.: '*The Design and Performance Analysis of QoS-Aware Edge-Router for High-Speed IP Optical Networks*', Proc of the London Communication Symposium, London (September 2000).

9   Crovella, M. E. and Bestavros, A.: '*Self-Similarity in World Wide Web Traffic: Evidence and Possible Causes*', IEEE/ACM Transactions on Networking, **5**(6), pp 835-846 (December 1997).

10

THE ROLE OF DEVELOPMENT IN COMPUTATIONAL SYSTEMS

R Tateson

10.1 Introduction

The development of a multicellular adult organism from the single egg cell is an amazing process. The complexity of the adult is vastly greater than that of the fertilised egg — there are many cells instead of just one, these cells display a range of different behaviours, and yet there is a co-ordination of behaviour to produce an identifiable unified entity, the organism, rather than an unco-ordinated soup of cells. But no new information has been provided during the course of development to allow the creation of complexity from (relative) simplicity.

This self-organisation of relatively simple parts into a complex and functional whole is a process which could be exploited for artificial systems. Computational methods, based on biological development, can be applied to a range of problems and would be particularly well suited to problems which are sufficiently large, complex, unpredictable and distributed where attempts to exert explicit central control are unsatisfactory. In such cases, the ability of local, autonomous 'cells' to continue to function and build an acceptable 'multicellular' whole in the absence of global knowledge can ease or remove the burden of central control. There are many telecommunications problems of this type which represent potential applications of developmentally inspired computing. Network growth, routing of data across networks, control of software agents and their environments, and allocation of resources (hardware, software or human) to maintain system performance could all benefit from technology which is, at some level, inspired by development.

10.1.1 No New Information?

There are two points which immediately suggest themselves from the paragraph above, and should be dealt with straight away. The first is an objection to the claim that no new information has been provided to allow the development of the adult. During development to adulthood new information which was not present in the fertilised egg is acquired, and in some sense literally 'incorporated'. This includes the whole spectrum of what might be called 'learning' — ranging from conscious efforts to gain knowledge about the world through to the unconscious tuning of senses and muscles to the physical reality in which the organism is immersed (for example, the existence of a gravitational force of a certain magnitude). However, this chapter is concerned with 'embryogenesis' — the proliferation and differentiation of cells and the interactions between those cells which allow the production of a meaningful creature even in the absence of significant outside influence. Consider the creation of a chick. This is a good example because the (huge) single egg cell with which we are all familiar is separate from the mother and can be successfully incubated in a variety of environments, requiring only warmth and air.

Everyone would agree that the chick which hatches from an egg, although it still has a lot to learn, is hugely more complex than the egg from which it originated. This is the case regardless of any efforts to provide or withhold sources of extra 'information'.

10.1.2   Origin of Existing Information

This leads to the second point. If one accepts that no new information has arrived, the question is: 'Where is this information hidden in the egg?' This has been a recurring question throughout the history of developmental biology. Many might presume that this question has been answered with the identification at the start of the 20th century of the nucleic acid as the chemical stuff of Mendel's hereditary mechanism. This has undoubtedly been a crucial realisation and led to explosive growth in knowledge and understanding of genetics throughout the 20th century. Indeed there is a beautiful symmetry to the discoveries of that century, with the initial finding that inheritance was by DNA (although it was not called that at the time), then the elucidation of the structure of DNA in the middle of the century which triggered huge strides in genetic understanding at the molecular level. Finally the century finished with a mass of information about the exact DNA sequence in the entire genomes of several species. So, is the answer to the question above 'DNA'; is the information which leads from egg to adult hidden in the DNA? In a sense it is. It is true that DNA is the stuff of inheritance and it encodes all the proteins which constitute the cells. However, the 'decoding' of the DNA into a multicellular animal is extremely involved and complex. The information for constructing an organism resides in the interactions between the DNA and its environment. DNA is interpreted by proteins to make proteins. These proteins in turn can directly or indirectly act on the DNA to change the ratios of proteins being made.

10.1.3 From Development to Computation

The unravelling of the deeply hidden DNA information to create a functional organism is the essence of developmental biology. How might this be applied outside the biological sphere? There are several levels at which developmental biology could be exploited.

The first is to select a particular, and fairly limited, instance of this natural 'decoding' or 'unravelling of complexity' and apply it by analogy to a problem in a different sphere. Providing the new problem shares sufficient similarities with the natural situation, the strengths of the natural process can lead to a similarly elegant solution in the new sphere. The analogy need not be rigidly adhered to, and may ultimately be abandoned as the technique is tuned to its new domain of operation. The first example given below (section 10.2) is at this level.

The second level of exploitation requires a broader analogy with a principle of the developmental process, rather than with a particular instance. In this case, one or more of the general features of developmental processes are applied to new problems. This level applies to the second example below (section 10.3), and particularly to the third example (section 10.4).

The third level is that of deliberate simulation. An attempt is made to capture important elements of natural development in some kind of abstracted system or 'model'.

There is a final level beyond this, which is not touched on by any of the examples below and is currently classified as science fiction. This is the literal connection of biological development with the new problem sphere. In this case one would be growing solutions. At successive levels it becomes less likely that the developmental inspiration behind it can ultimately be cast off without destroying the value of the solutions.

It also becomes increasingly likely, moving through the levels, that the developmentally inspired solution will be dynamic and linked in some way to the new problem domain. In other words whereas one could imagine a first level solution which was used once to design a system, but was never 'online' itself, this becomes less likely with a simulation. In that case one would expect the 'solutions' to be continuously updated and the simulation would be either 'inhabiting' an environment containing 'problems' or would itself constitute such an environment.

10.1.4 Previous Work

There has been a wealth of work on the third level of developmental 'exploitation' as outlined in the previous section — that is as a simulation or model of natural systems. Much of this has been done by biologists attempting to test theories concerning the working of their experimental systems —communities, individuals, organs, tissues or cells. There have also been many contributions from biological theorists and computer scientists [1-4]. Work on modelling neural growth is of particular relevance because it can combine the wish to model natural phenomena accurately with the hope of applying those models to artificial design. The work of Vaario et al [5] uses biologically plausible models of neural outgrowth and seeks to apply the results to a better understanding of nature and improvements in the design of control systems for robots. Rust et al [6] employ a different, but also biologically realistic, approach to modelling neural growth. In this case it is the activity-based pruning of the network which produces efficient results via a compact developmental encoding.

10.1.5   Benefits of Developmental Computation

The final question to address before moving on to give some examples of this kind of computational exploitation of development is: 'What are the grounds for believing that these approaches will be stronger thanks to their developmental inspiration?' In fact each new approach must be able to justify itself in the new problem domain. No matter how elegant the natural mechanism, and no matter how perfect the computational mimicry, it will be judged a failure if it can offer no actual or potential advantages over existing approaches to the problem of interest. However, there are reasons for optimism that techniques based on developmental biology are likely to provide such advantages in certain classes of problem domain.

Most importantly, the cellular nature of biological organisms is tightly bound up with their development. The mechanisms at work are such as to allow a highly complex and stable system (the functional organism) to emerge from the autonomous actions of a large number of cells each carrying the same fundamental instruction set (the DNA sequence of the organism in question). It is therefore likely that developmental mechanisms will preserve their natural advantages if they are applied to computational problems for which the cellular analogy can be made.

A second feature of natural development is its use of space. Cells modify their environment, locally or at a distance, thereby influencing their neighbours. This is hardly remarkable in natural systems. Indeed, it would be surprising if developmental mechanisms which produce such exquisite three-dimensional detail could do so without using spatial signals. Computational systems, on the other hand, need not be tied to the spatial dimensions of our world. Solutions based on developmental analogies are likely to be most applicable when there is a spatial aspect to the computational problem. This could be a direct spatial context, such as the physical layout of a network, or some more abstract spatial context such as logical proximity of network nodes.

The final point, which flows naturally from the observations that developmental systems are cellular and spatial, is the distributed manner of developmental problem solving. The outcome is not directed by a central controller, but emerges from the multitude of local decisions made by cells. This has advantages for natural organisms because it allows recovery of 'correct' patterns through local processes (analogous to healing in the adult) without the necessity for a central planner to be 'notified' so that a new plan can be produced. For example, if a neuron in a developing fruitfly dies, its place can be taken by a neighbouring cell which then develops to form a neuron [7]. This cell would otherwise have been epidermal (making 'skin' not 'nerve'), but, since there is spare 'skin-making' capacity, its absence will not be missed. The absence of a neuron, on the other hand, would reduce the ability of the fly to sense and respond to its world. If this process of developmental pattern repair was to be orchestrated by a central controller, it would require messages concerning the current state of the pattern to be sent from all parts of the organism. The controller would then have to deliberate in some way and then disseminate instructions to the appropriate regions. Distributed problem solving can also be advantageous in many artificial systems. Any problem domain where the solution must apply to a disperse and rapidly changing 'world' is likely to benefit from distributed processing. Such problems will be in increasingly plentiful supply as telecommunications and computation networks continue growing in size and complexity, defying central control and requiring instead distributed solutions.

10.2   Retinotectal Pathfinding

Light entering the human eye stimulates the rod and cone cells of the retina. The resulting signals from these cells pass through two further layers of processing cells before leaving the eye via the optic nerve. After bearing right or left at the optic chiasm (the crossroads at which neurons in the optic nerve may cross to the other side of the brain) the signals arrive at the optic tectum (see Fig 10.1). This is the first stage at which visual signals are integrated with information from other senses, and it is the last stage at which a one-to-one mapping is maintained between the pattern of light entering the eye, and the physical distribution of neurons in the brain.

The correct routing of the signals during adult life depends on the correct routing of the neurons which will carry those signals. These neurons grow during development from the retina towards their targets in the tectum (see, for example, Dowling [8] and Purves et al [9] for a more detailed explanation).

10.2.1   Pathfinding Mechanisms

Several mechanisms are used by the neurons in their growth towards the tectum. These are based on the fact that at any moment a certain neuron has a certain complement of receptors on its surface. Receptors are protein molecules synthesised by the cell, which are generally placed in the cell membrane such that one end is exposed to the environment outside the cell, while the other end is in contact with the 'cytoplasm' inside the cell. The external part of the receptor binds particular molecules, known as ligands, with great specificity. Ligand specificity varies from one type of receptor to another. Binding of a ligand to a receptor can have a wide range of effects from simple physical adhesion to a complex change in the behaviour of the cell. Different types of ligands can be produced by the cells around the neuron and may remain anchored to the producing cell, necessitating a direct contact with the neuron to allow receptor-ligand binding, or may be free to diffuse away from the source, creating a concentration gradient which might attract or repel the neuron. This allows a subtle and multifactorial interaction between the neurons and the cells on the route to the targets.

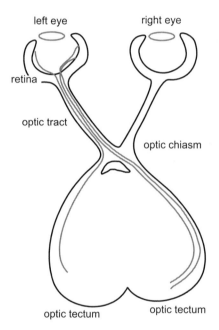

Fig 10.1   The path followed by neurons growing back from the retina to their targets in the optic tectum. The figure shows a diagrammatic cross-section of the visual pathways in the brain of a vertebrate (seen from above). Neuron growth is shown from the left retina, along the optic tract to the optic chiasm where neurons must decide whether to cross to the other side of the brain, then onward to targets in the tectum.

Examples include:

- fixed ligands on the route to the target which are bound by receptors on the neuron, allowing the neuron to 'crawl' towards the target;

- concentration gradients of diffusible ligands which can be followed by neurons with the appropriate receptors;

- temporally varying expression/regulation of both receptors and ligands allows different combinations over time;

- spatially varying expression of receptors and ligands, again allows neurons and other cells to interact in a huge variety of different ways, dependent in this case on position;

- activation of receptors by ligands — in addition to physical binding between ligand and receptor, the act of binding can trigger behaviour changes in the receptor cell.

It is through these types of control mechanisms that the retinotectal system can maintain spatial order and an inverted image on the tectum that is the same as that perceived by the retina. This 'retinotopy' allows complex information to be dealt with in an efficient way while reducing confusion between tens of thousands of growing axons during development. It is impossible that a genetic blueprint dictates the number of different axons, connections that underlie formation of the retinotectal pathway, since this would require tens of thousands of different genes. Instead regulating the spatial and temporal expression of a particular set of genes can provide many different types of interaction that may govern both the modularity and specificity during nervous system development.

10.2.2   Simulation

These mechanisms have been used as the basis for a highly abstracted simulation of retinotectal pathfinding which has potential as a useful connection generator for artificial networks (see Fig 10.2). A simple artificial neural network model was created in which each 'neuron' could in principle make a connection to any other neuron. Hand design of the connections allows the network to perform rudimentary information processing, but for the purposes of the retinotectal simulation the aim was to use a developmental process to make connections from the primary layer of neurons (the retina) to designated targets in subsequent layers. Rather than providing the retinal neurons with co-ordinates of their targets, they are 'born' with a set of receptors (different combinations for different neurons) and a set of simple intrinsic growth behaviours (identical for all neurons). The neurons over which the retinal neurons must grow to reach their targets carry various ligands, and the target region has a gradient of a different ligand. By following their growth rules, in the context of their own complement of receptors and the ligands in their environment, the neurons can seek out the correct targets.

10.2.3   Applicability

In simulation the developmental mechanisms have been used to configure a neural network which is simple, static, and entirely under the control of the user. Hence

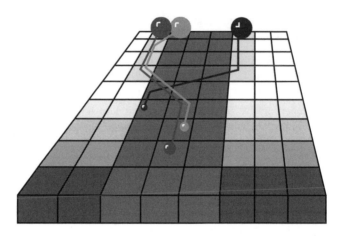

Fig 10.2 A simulation of the retinotectal system. Neurons originate at the top of the diagram (corresponding to the retina) and are given a small set of simple rules (corresponding to the receptors they would express). Their growth is then directed by the interaction between these 'receptors' and the 'ligands' expressed by the cells of the underlying medium. 'Ligand' expression is shown by the various densities of shading in the square 'cells'.

there is no benefit to using a developmental approach. It would be easy to simply provide co-ordinates for the connections. This kind of approach is only going to reap dividends in complex networks in which targets shift rapidly and unpredictably and where the user does not have control of the whole system. In such cases it may be easier to 'design' a type of 'neuron' with a set of 'receptors' and allow it to find its own way to a target than to attempt to explicitly specify the position of that target.

10.3   Notch/Delta Signalling

Developing cells carry on their surfaces a large number of different types of signal and receptor molecules which allow them to communicate with neighbouring cells. A given type of signal molecule will bind specifically to one, or a few, types of receptor. As a consequence of signal binding to receptor, a message is usually passed into the cell on whose surface the receptor is carried.

One example of such a pair of signal and receptor is Delta (the signal) and Notch (the receptor) — Fig 10.3 shows a diagrammatic description of the Notch/Delta signal. Versions of these molecules are found in many different animals, from nematodes (*C elegans*) to humans (*H sapiens*). They are used for communication between cells at many stages in development. For example, in the fruitfly Drosophila, they are needed for correct nervous system formation in the early

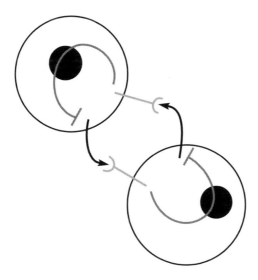

Fig 10.3 Diagrammatic description of the Notch/Delta signal. Two cells are shown, each producing a signal (Delta shown by the darker arrow) and a receptor (Notch shown by the lighter fork). As a result of Delta binding to Notch an inhibitory signal is sent (blunt arrow) which reduces the strength of Delta signalling by that cell. In the two-cell case shown here, the result is a bistable switch with one unstable equilibrium (the cells produce exactly equal amounts of Delta and Notch) and two stable equilibria (upper cell gains upper hand, increases Delta production and forces lower cell to cease production, or vice versa).

embryo and for developing the correct wing shape during pupation [10-12]. Figure 10.4 shows the stages of development of cells acquiring the potential to make bristles over a 12-hour period.

Although there are other molecules which can have effects on the communication between Delta and Notch, the core of their interaction is simple. The binding of Delta to Notch causes a protein to enter the nucleus of the Notch-carrying cell and alter the state of expression of the DNA. The effect of the alteration is that the cell's production of Delta is reduced. Thus its ability to send a signal like the one it has just received is diminished. Production of Delta is linked to a choice of cell type. In other words the cells having this conversation are 'contemplating' becoming one sort of cell (for example neural), which correlates with Delta production, rather than another (for example epidermal). The exact nature of the choice depends on the particular developmental process at hand; the Delta and Notch molecules are not concerned with that, just with ensuring that a choice is made. So, if a cell perceives a Delta signal from another cell, it makes less Delta itself and hence becomes both less able to signal Delta to its neighbours, and less likely to choose the Delta-correlated cell type.

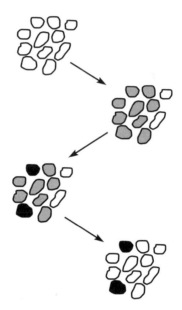

Fig 10.4   In the developing fruitfly the Delta/Notch signalling system is used in a relatively large, two-dimensional field of cells. It allows the production of an acceptable pattern of bristles. The figure shows the stages of development occurring over the course of 12 hours. Initially, many cells acquire the potential to make bristles (first arrow, bristle potential shown in grey). Next (second arrow) the Delta/Notch 'negotiation' begins among those grey cells resulting in some getting darker (more likely to make bristles) and inhibiting their neighbours. Finally (third arrow) the process terminates with a few cells determined to make bristles and the rest forming the surrounding exoskeleton.

10.3.1   Frequency Allocation

This developmental mechanism was the inspiration for a new method for solving a well-known problem in mobile telephony networks. Known as the frequency allocation problem, it arises from the concept of frequency reuse. The limited number of radio frequencies available to the operator of a mobile telephone network must be allocated to the much larger number of base-stations in the network such that those base-stations have enough frequencies to meet demand for calls without reusing a frequency already being used by another nearby base-station.

The method [13] relies on the analogy between cells and base-stations. Each base-station is made to negotiate for use of each of the frequencies, just as the cells negotiate for cell fate choice. For every frequency, a cell has a simulated 'Notch receptor' and is synthesising simulated 'Delta' in an effort to inhibit its neighbours. All base-stations begin with almost equal preference for all frequencies. The small inequalities are due to the presence of noise. Over the course of time, as the negotiations continue, they abandon most frequencies (due to inhibition by their neighbours) while increasing their 'preference' for a few frequencies.

This approach has been tested in a simulated network as shown in Fig 10.5. The benefits of the method are exactly those which would be anticipated due to its developmental inspiration. It provides dynamic, robust solutions which continue to meet the demand for calls even as traffic fluctuates in the network. Because the method is inherently distributed, placing the processing load at the base-stations rather than at a central controller, it is able to run continuously online and its performance does not suffer as the network expands.

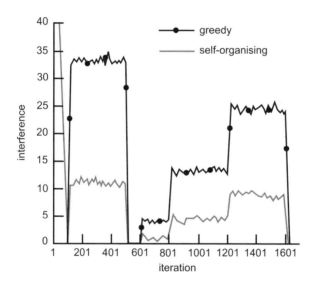

Fig 10.5   Performance of the proposed method in a network simulation. The horizontal axis shows iterations of the algorithm — equivalent to time. The vertical axis shows predicted interference (on an arbitrary scale) in the whole network. This value should be minimised to optimise quality of service. Having produced an optimal frequency allocation plan (by iteration 100) an increase in demand is simulated at some of the nodes in the network. Demand drops again at iteration 500. Demand again increases, in 3 steps corresponding to the involvement of an increasing number of nodes, from iteration 600. Current frequency allocation methods would block the excess calls so interference would not change, but customers would be unable to make calls. Therefore, for meaningful comparison the proposed self-organising method based on Delta/Notch signalling is plotted alongside an alternative dynamic algorithm ('greedy') which does not do any negotiation between nodes, but blindly assigns sufficient unused frequencies such that demand can be met.

10.3.2   Applicability

The method inspired by development differs markedly from current practice in frequency allocation. Currently the network operators use centralised planning — at regular intervals a planning committee meets to decide on a new frequency plan (according to network performance measures since the last such meeting). This plan is then disseminated to the base-stations, and will remain in place until the next meeting.

The current method works best when networks are relatively small and there is little fluctuation in traffic. The larger the network, and the greater the fluctuations, the harder it becomes for a centralised solution strategy to continue to deliver acceptable quality of service. In effect, the efficiency with which the resource (frequencies) is being used declines because that resource cannot be moved to meet demand. As a result the operator must invest in constructing new base-stations (i.e. buying more resource) to maintain quality of service. It is in exactly these circumstances of dynamic traffic in a large network that the method inspired by development is likely to deliver tangible benefits.

10.4   Dynamic Genome

The final example is in the spirit of simulation, rather than application. The aim is to simulate the essential features of cells such that over the course of the simulation those cells can 'evolve' new and advantageous behaviours. Ultimately it is hoped that this simulation will include multicellular behaviour such that groups of cells will divide a larger task between them.

Development, indeed all the life of the majority of cells, is orchestrated by the interaction between the DNA and the environment. The environment includes the interior of the cell — the cytoplasm — which can act almost directly on the DNA, and the exterior of the cell — the surface, and the immediate surroundings — which can only act on the DNA indirectly.

The DNA encodes a large number of proteins, and almost every cell in a multicellular creature has an entire copy of the DNA. Hence all cells have the potential to make all proteins. Of course they do not actually synthesise them all. Those proteins which are currently being synthesised, and the levels at which that synthesis is taking place, constitute the current 'expression state' of the cell. It is the expression state to which people are usually referring when they talk of 'cell type'. So, for example, neurons (nerve cells) and lymphocytes (immune system cells) are different cell types. They have become different over the course of development, and so their history is very important for determining their cell type, but even in the absence of any information about their origins, it is possible to tell neurons from lymphocytes purely on the basis of their current expression state.

Until very recently expression state would be sampled by testing for the presence or absence of a few proteins which would unambiguously distinguish between cell types. It is now possible to look at a much broader range of gene activity — levels of expression of many hundreds of proteins can be measured in a single cell.

Over the course of development, each cell moves through 'expression state space' as well as the physical space of the organism of which it is a part. The expression state defines the cell moment by moment, and constantly re-creates the cell by synthesising the proteins, the cell's physical constituents and catalysts.

How best to capture the navigation through expression state space? One approach, described in detail in Hoyle and Tateson [14], is to simulate a cell with a range of behaviours including division, movement and synthesis of 'chemicals'. This cell responds to its chemical environment by selecting behaviours, which can modify the chemical environment and produce more cells with identical 'DNA'. An alternative approach, described below, focuses on the dynamic regulation of the genome.

10.4.1   Simulated Cells with a Two-Way, Dynamic, Genome-to-Phenotype Mapping

There are many artificial life systems which simulate populations of relatively simple 'creatures'. These have proved useful for demonstrating the emergence of complex behaviours at a level above the individual (e.g. birds flocking [15], wasps nesting [16], and ants forming trails [17]). But such models do not attempt to give insights into the internal workings of the individual. The proposal offered here is to drop the level of description down one level. The individuals are simulated bacteria and their behaviour will be analysed at the level of the individual.

* Why?

  A bacterium is a highly evolved and effective agent in a complex and uncooperative world. Searching for organisational principles within a bacterium is likely to provide insights into good design for multi-agent, adaptive systems. Hopefully it will also be a useful tool for testing models of real bacterial function, e.g. bacterial chemotaxis — the movement of a bacterium towards a source of attractant by swimming up the concentration gradient created by diffusion of attractant away from the source.

* How?

  The aim is to simulate the web of interactions which connect the genome with the external environment. In contrast to existing, higher level simulators, all levels of this web must be dynamic. Thus, the genome is not decoded into a phenotype which then interacts with the world. Rather the expression of the genome depends on the internal environment of the cell. The internal environment depends on the history of the cell, the current expression of the genome and the external environment. The impression which the external environment makes on the internal environment depends on the receptor proteins in the membrane of the cell (which, in turn, depend on the expression of these proteins by the genome).

10.4.2   Overview of a Real, Generalised Bacterium

A bacterium is a single-celled organism. It is typically smaller than a eukaryotic cell (either free-living or in multicellular organisms) and its internal structure is simpler [18]. It has a single, circular chromosome carrying almost all the genes it might use during its life. In addition, it may have one or more small extra chromosomes carrying genes for things like antibiotic resistance or mating type. The cell is surrounded by a lipid membrane which is impermeable to all but the tiniest molecules (e.g. oxygen can diffuse straight through but glucose cannot). The lipids forming the membrane are molecules with a hydrophilic ('water loving') end and a hydrophobic ('water hating') end. In an aqueous environment these molecules tend to form layers two molecules thick with the hydrophobic ends of all the lipids sandwiched in the middle and the hydrophilic ends exposed to the watery environment. The cell membrane is an example of such a 'lipid bilayer' which has closed on itself to form a sphere. The membrane does not provide any structural rigidity so outside it there is a cell wall made of a cross-linked polymer. The need for structural strength arises from the fact that the contents of the bacterial cell are usually more concentrated than the external environment.

This results in water being drawn in by osmosis and can lead to cells bursting. The strength provided by the cell wall allows the cell to maintain a high osmolarity of its cytoplasm relative to the outside world without bursting due to the pressure (in other words it can have a rich soup of stuff inside even though it is floating in a very thin soup environment). The bacterium's genes encode the proteins and RNA molecules it needs to carry out all its functions — moving, gathering food, breaking food down, synthesising new material, reproducing. Appropriate control of these functions, dependent on internal and external conditions, is achieved by multi-layered, cross-connected control networks. Short-term, rapid responses usually rely on existing structures to cascade signals and trigger appropriate action. Longer term, slower responses may involve synthesis of new material, possibly by changing the expression state of the genome. For example, bacteria are able to respond to changes of concentration of food molecules in their environment using a rapid response. The control network involves receptors at the surface signalling to motors via a relatively short chain of messenger proteins. As genes need to be turned on or off, the signal need not involve the genome. By contrast, to adapt to a change in the nature of the food (e.g. lactose instead of glucose) the cell must synthesise new enzymes. The signalling cascade in this case includes the genome and alters its expression state — genes encoding proteins for glucose usage are switched off and those encoding proteins for lactose usage are switched on.

10.4.3   Simulation

'CellSim' is the simulator of the proposed cellular architecture. It comprises 100 'cells'. Each cell has a 400-bit genome. This genome is divided into 10 genes, each with 40 bits. Each gene is divided into 5 domains, each with 8 bits. These domains are called 'more', 'less', 'func', 'modi' and 'act'. Figure 10.6 shows the genome structure of the simulated cells.

The interior of each cell is simulated as a 'well-stirred soup' of binary strings. These strings are able to bind each other, and by so doing will affect the behaviour of each other (and ultimately the behaviour of the cell at a gross level). When a product is made from a gene, its sequences (specified by the 'modi' and 'act' domains) are added to the soup and are able to bind to other strings.

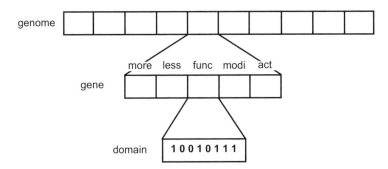

Fig 10.6   The genome structure of the simulated cells. The genome consists of 10 'genes'. Each gene has 5 domains: more, less, func, modi and act. The 'more' and 'less' domains determine the 'expression level' of the gene, i.e. the rate of production from that gene. The 'func' domain dictates the nature of the product, in terms of which compartment of the cell it will occupy. 'modi' and 'act' specify the sequence of the product. Each domain is encoded by 8 binary bits.

A 'soft matching' process is implemented, allowing the simulation of complementary binding between bit strings. For example the string 11110000 would bind with highest affinity to the string 00001111, and with lowest affinity to another 11110000 string.

It is the binding of strings already in the soup to the 'more' and 'less' domains which decides the level of expression of the gene. The 'occupancy' of these domains is the sum of the binding affinities of all soup strings to the domain — the greater the 'occupancy' of the 'more' domain, the greater the expression level; conversely, the greater the occupancy of the 'less' domain, the lower the expression level. Figure 10.7 illustrates gene expression.

In addition to binding to the gene domains, the soup strings can bind to each other.

The 'act' sequence can bind to the 'modi' sequence of other strings, and hence alter the function of the string. The nature of the function is determined by the 'func' domain, but the level of activity varies depending on 'occupancy' of the 'modi' sequence.

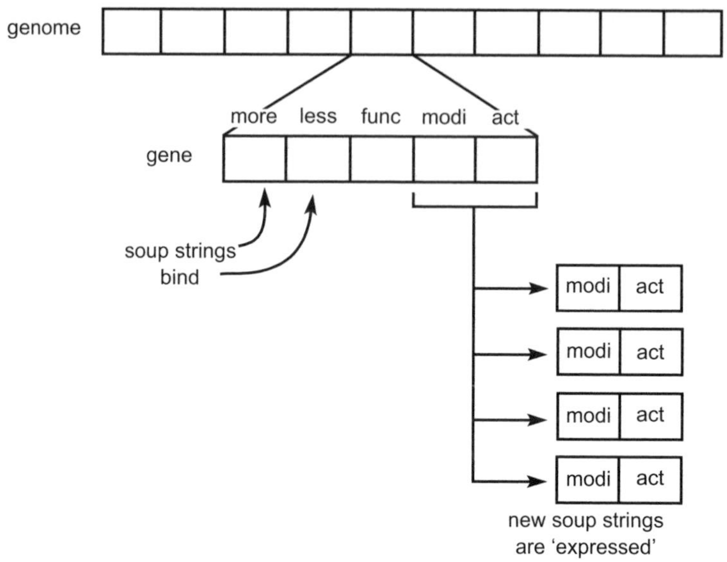

Fig 10.7   Gene expression. Strings already synthesised can 'bind' to the 'more' and 'less' domains to affect level of expression of the corresponding 'func/modi/act' string.

The simulation iterates so the new additions to the mixture of soup strings immediately have the opportunity to bind to the gene domains and all the other soup strings.

This internal feedback loop is linked to the world outside the cell in two senses. Firstly, the outside world can impinge on the soup, either by adding strings directly to the soup (analogous to chemicals diffusing into the cell) or by providing strings which can bind to soup strings which have their 'modi' sequence sticking out of the cell, and hence available to interact directly with external strings. Secondly the cell can impinge on the outside world, either by emitting strings into the external environment, or by performing some activity (such as moving in a spatial world). Activities like this are defined by the 'func' domain.

10.4.4   Results and Applicability

To date the simulated cells have been tested in an extremely simple 'world' containing only their siblings and a food supply (see Fig 10.8). Each cell was provided with a 'motor' which it could switch on or off using the same soft matching process used for gene expression and protein interaction. In addition, each cell had a receptor with a binding site which matched the food. The implicit task for the cells was to identify a control loop which would link that receptor to the motor in such a way that the cell was able to 'swim' up a concentration gradient of food. This was an implicit task because the cells were not evaluated directly according to their ability to perform the task. Rather they were only allowed to reproduce when they had absorbed a certain amount of food. Cells which could climb a gradient tended to accumulate food faster and were hence 'rewarded' with reproduction. Over generations, cells appeared which were capable of gradient climbing.

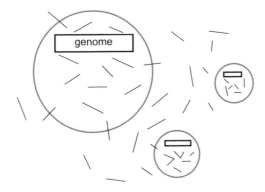

Fig 10.8   Cells in a CellSim world.

One anticipated application of the cell simulation is in the future of software agents. In a situation where a complex network environment exists, populated with a range of software agents, the ability to evolve to perform tasks may become useful. This is most likely to be the case when some, or all, of the agents in the environment are relatively simple and are performing high-level tasks as a group rather than as an individual. In these circumstances of emergent problem solving, there will be niches for soft-matching evolving 'cells' to exploit.

CellSim could also find application in a future of 'active networks' in which the nodes within a telecommunications network provide services rather than merely routing requests to the appropriate server. A cell-based approach to modelling the functionality of active nodes has been proposed based on bacterial gene transfer [19]. This technique shows advantages in terms of allowing the network as a whole to improve its ability to meet service requests by the local actions of the 'bacterial' nodes.

10.5   Discussion and Summary

This chapter has described three examples of developmental biology influencing novel computation methods. The first example, retinotectal pathfinding, is an example of strong biology with potential for future application to 'growing' telecommunications networks or artificial neural networks. The second example,

Notch/Delta signalling, has found a direct application which preserves elements of the biological specifics as well as the more fundamental cellular basis. The third and final example does not seek to address the specifics of any one developmental process, but aims to simulate the sort of 'computing' interactions common to all cells during the processes of development and differentiation. It may find application as a 'soft computing' technique for engineering adaptive bottom-up computational architectures.

In all cases of developmentally inspired computation there are scenarios which will tend to favour such approaches over existing methods. An important challenge for future work is to identify such scenarios and bring these novel techniques to bear in tackling these problems.

References

1 Eggenberger, P.: '*Evolving morphologies of simulated 3D organisms based on differential gene expression*', in Husbands, P. and Harvey, I. (Eds).: '*Artificial Life*', Proceedings of Fourth European Conference, MIT Press, pp 205-213 (1997).

2 Meinhardt, H.: '*Models of Biological Pattern Formation*', Academic Press, London (1982).

3 Savill, N. J. and Hogeweg, P.: '*Modelling morphogenesis: from single cells to crawling slugs*', J Theor Biol, **184**, pp 229-235 (1997).

4 Turing, A.: '*The chemical basis of morphogenesis*', Phil Trans R Soc, **B237**, pp 37-72 (1952).

5 Vaario, J., Onitsuka, A. and Shimohara, K.: '*Formation of neural structures*', in Husbands, P. and Harvey, H. (Eds): '*Artical Life*', Proceedings of Fourth European Conference, MIT Press, pp 214-223 (1997).

6 Rust, A. G., Adams, R., George, S. and Bolouri, H.: '*Activity-based pruning in developmental artificial neural networks*', in Husbands, P. and Harvey, H. (Eds): '*Artificial Life*', Proceedings of Fourth European Conference, MIT Press, pp 224-233 (1997).

7 Stern, C.: '*Two or three bristles*', Am Scientist, **42**, pp 213-247 (1954).

8 Dowling, J. E. '*Neurons and Networks — An Introduction to Neuroscience*', Belknap Press of Harvard University Press, Cambridge, MA (1992).

9 Purves, D., Augustine, G. J., Fitzpatrick, D., Katz, L. C., LaMantia, A-S. and McNamara, J. O. (Eds): '*Neuroscience*', Sinauer Associates, Sunderland, MA (1997).

10 Artavanis, T. S., Matsuno, K. and Fortini, M. E.: '*Notch signalling*', Science, **268**, pp 225-232 (1995).

11 Fehon, R. G., Kooh, P. J., Rebay, I., Regan, C. L., Xu, T., Muskavitch, M. and Artavanis, T. S.: '*Molecular interactions between the protein products of the neurogenic loci Notch and Delta: two EGF-homologous genes in Drosophila*', Cell, **61**, pp 523-534 (1990).

12 Heitzler, P. and Simpson, P.: '*The choice of cell fate in the epidermis of Drosophila*', Cell, **64**, pp 1083-1092 (1991).

13 Tateson, R.: '*Self-organising pattern formation: fruit flies and cell phones*', in Eiben, A. E., Bäck, T., Schoenauer, M. and Schwefel, H. P. (Eds): '*Proceedings of 5th International Conference*', PPSN, Springer, Berlin, pp 732-741 (1998).

14 Hoile, C. and Tateson, R.: '*Design by morphogenesis*', BT Technol J, **18**(4), pp 112-121 (October 2000).

15 Reynolds, C. W.: '*Flocks, herds, and schools: a distributed behaviour model*', Computer Graphics, **21**(4), pp 25-34 (1987).

16 Theraulaz, G. and Bonabeau, E.: '*Coordination in distributed building*', Science, **269**, pp 686-688 (1995).

17 Dorigo, M., Maniezzo, V. and Colorni, A.: '*The ant system: optimization by a colony of cooperating agents*', IEEE Transactions on Systems, Man and Cybernetics — Part B, **26**(1), pp 29-41 (1996).

18 Alberts, B., Bray, D., Lewis, J., Raff, M., Roberts, K. and Watson, J. D.: '*Molecular Biology of the Cell*', (3rd Edition) Garland Publishing, New York (1994).

19 Marshall, I. W. and Roadknight, C.: '*Adaptive management of an active service network*', BT Technol J, **18**(4), pp 78-84 (October 2000).

11

ADAPTIVE SECURITY AND ROBUST NETWORKS

F Saffre

11.1   Introduction

In the last ten years, using biology as a source of inspiration has become widespread in long-term research for security in information technology. The reason is that problems faced by distributed computer systems when under attack are reminiscent of those that their biological counterparts have had to deal with (and find solutions for) throughout their evolution. Indeed, pathogens attempting to invade an organism, or a population of organisms, have probably been around from the very beginning of life on earth, and with them, the vital necessity to limit the damage they cause.

This can be done in two different ways — either by keeping the 'invader' in check at the individual level, or by preventing it from compromising the survival of the entire population. Similarly, two main currents can be distinguished in bio-inspired IT security — one focused on the microscopic level (immune response), the other on macroscopic dynamics (epidemic control).

An illustration of the microscopic approach can be found in the seminal work by Stephanie Forrest et al [1] on artificial immune systems. The primary aim of this research is to devise strategies to successfully and efficiently discriminate between the equivalents of 'self' and 'non-self' in very large sets of heterogeneous data. This would typically be achieved by 'training' a population of 'antigens' that would only match bit-strings that are totally absent from uncontaminated files. The main advantage of this approach is that, in theory, it would be able to identify unforeseen modifications, thereby pointing to any file that has been corrupted, even by an unknown virus. However, to date this promising work has failed to deliver practical security solutions, probably because of scalability and flexibility issues.

The macroscopic approach on the other hand, far from assimilating the network to one individual protected by a single immune system, regards it as a population of interacting entities. Its primary focus is therefore on the dynamics of contagion/

recovery and its long-term objective is to devise better strategies for epidemic control. One illustration of this approach would be the investigation by Kephart and White [2] of the influence of topology on the spreading of computer viruses, pointing out similarities with more generic population dynamics patterns [3]. Although this work did not receive the attention it deserved when first published, it can be argued that it is about to be spectacularly revived as networks become more dynamic and so do the threats (pathogens, cascade failure, etc) menacing them.

Our vision of bio-inspired IT security is located somewhere in between these two approaches. We start from the hypothesis that a pluricellular organism can be assimilated to a population of individual cells that have evolved a set of mutually beneficial interactions. Similarly, large distributed architectures can be regarded as symbiotic associations of otherwise isolated computers. In both cases, in order for sub-units to identify each other as members of the same community, there is an obvious need for an 'inclusive' sense of 'self', which in the biological world translates into a phenomenon known as histocompatibility. This basically means that normal cells do not trigger a 'hostile' reaction when inspected by agents of the immune system, because they display the right 'ID tag', in the form of specific molecules on the membrane. Likewise, the first step toward protecting a computer network is to devise a way of discriminating between the many benign interactions involving only legitimate members of the community on both ends, and the odd intrusion attempt emanating from what is actually the entry point of a malevolent entity.

So far, the analogy we propose is very close to the artificial immune system (AIS) paradigm. Yet instead of adopting a traditional 'mean-field' approach, we suggest that the spatial relationships between individual sub-units can be exploited in order to identify which are more likely to be targeted by assaults from the outside world. These 'peripheral cells' can then be encouraged to 'differentiate' and special-ise in carrying out security procedures, forming a protective layer around other members, which can therefore focus on performing other tasks. As a matter of fact, it is precisely this idea, that a well-defined interface exists between the 'inside' and 'outside' worlds, that underlines the firewall concept (i.e. the backbone of contemporary network security strategy). However, because of the rapidly increasing plasticity and mobility of constitutive devices, network boundaries have become extremely fuzzy and changing, effectively requiring constant re-deployment of defensive measures. In the IT security community, this problem has become a very serious concern, sometimes provocatively referred to as 'the disappearing perimeter' [4].

One of the aims of the work described in this chapter is to demonstrate that any adaptive behaviour capable of successfully dealing with such a dynamic threat would in principle closely resemble the inflammatory response of the mammalian immune system, whereby lymphocytes rapidly accumulate in the region surrounding an opening breach. It is the inherently dynamic nature of this re-organisation process, which involves using variable spatial gradients to identify exposed areas, that links the present work to the macroscopic, epidemiological approach.

11.2   Epidemic Propagation and Cascade Failure

Many recent studies have underlined the critical influence of topology on network survivability. It has been shown [5-7] that different designs react differently to both random node failure and directed attack against key relays. In a closely related work, Pastor-Satorras and Vespigianni [8] demonstrate that the topology of the Internet actually favours the spreading of computer viruses. Finally, among other observations on the properties of complex networks, the review by Strogatz [9] emphasises the role of design on catastrophic cascade failures.

In order to determine to what extent adaptive security procedures can limit the amount of damage caused to a network by the propagation of a malicious entity, it is vital to establish a suitable benchmark. In particular, as one of the key issues addressed here revolves around the fact that topology is likely to become an increasingly dynamic real-time variable in distributed systems, the response of several possible designs should be carefully examined before robust solutions can be discussed. In this section, we investigate different scenarios, involving two unprotected network architectures. One is close to the scale-free topology, widely accepted as a realistic model of the Internet [10], the other is an original hybrid topology termed 'hypergrid', which we argue has desirable features in terms of robustness and scalability.

One of the most important things to understand when comparing topologies is that they dictate how traffic is distributed in the network. For example, in a scale-free architecture, the power-law distribution of node degree has clear implications in terms of how many packets transit through each relay in day-to-day operation. This in turn dictates how much capacity the corresponding hardware must have in order to maintain acceptable quality of service (a local router obviously does not require the same amount of bandwidth as one of the core relays handling intercontinental traffic). Because there are economic constraints, one ends up with a network where each node is perfectly capable of assuming its role in normal conditions, but also has a limited tolerance to brutal and/or unpredictable surges. The result is a potentially very volatile environment as, whenever a relay fails (or is targeted by a malicious entity), the traffic it previously supported typically requires re-routing. This in turn increases the workload on those other nodes which start acting as back-ups. In an extremely hierarchical architecture (such as a scale-free network), this can have catastrophic consequences if one of the main hubs ceases to function. Indeed, a huge amount of traffic may need to be transferred to secondary relays that simply lack the capacity to process it. If overload causes them to crash too, it can easily initiate a chain reaction, as more packets have to be re-routed through increasingly less capable nodes, resulting in the perfect example of cascade failure.

This has very serious implications for network survivability, as it basically means that the damage caused by a malicious entity targeting relays can extend far beyond those nodes that it successfully infected. Albert et al [5] take resilience of the largest component (largest sub-set of interconnected nodes) as the sole criterion for robustness and rightly conclude that scale-free networks are very resistant to random failure. Although a propagation of a virus cannot be reduced to a succession of such random events (because it typically follows a geometrical progression starting from a specific location), this could mistakenly be interpreted as an indication that the epidemic would not threaten the system as a whole if rapidly contained.

We would argue that, although coherence of its largest component is obviously a necessary condition if a network is to remain operational when submitted to stress, it is in fact not sufficient in the case of the Internet (or in fact any other distributed system where traffic is an issue). This can be illustrated by the results we obtained using Monte Carlo simulation of a medium-sized network.

11.2.1   Results for the Reinforced Scale-Free Network

We used the preferential attachment rule described by Barabási et al [11] to build a scale-free system comprising 1024 nodes, each newly added member (first excepted) initiating one undirected connection with the existing population. We then added an extra 1024 random links in order to increase the average degree to $k = 4$, generating a very robust design in terms of the resilience of the largest component (see Fig 11.1). To model traffic, we wrote routing tables on a simple shortest path basis, then simulated transmission of one packet from every node to every other (1024^2 transfers) and recorded node usage over the session (total number of packets that transit through each relay). This value is used as an indication of 'normal' load and we assume that resources are allocated accordingly. As can be expected, a node's utilisation is correlated with its degree — highly connected hubs typically ending up handling all traffic between distant regions of the network (Fig 11.2).

A total of ten networks, sharing those same global characteristics (number of nodes/links and degree distribution), but differing in terms of their local characteristics (distinct pseudo-random number sequences) are created following this procedure. For each of them, we then simulate an epidemic, starting from a single infected node. The malicious entity is assumed to be self-replicating at a constant rate of 0.1 — at every time-step, every first neighbour of an infected nodehas a 10% chance of being targeted (as there is no protection at this stage, every attempt is successful). All updates are conducted 'simultaneously' (network state at time $t + 1$ is computed from data collected at time t), so the invader can never propagate faster than one hop per time-step. We allow the epidemic to spread in that fashion for 16 time-steps, after which an average ~17% (SD ~7%) of the 1024 nodes are infected. As can be expected from the profile shown on Fig 11.1,

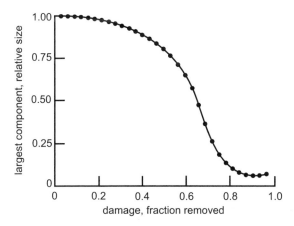

Fig 11.1   Evolutions of the relative size of the largest component in a reinforced scale-free network comprising 1024 nodes and 2048 links, when submitted to cumulative node failure (profile obtained using BT's own robustness analyser (RAn)).

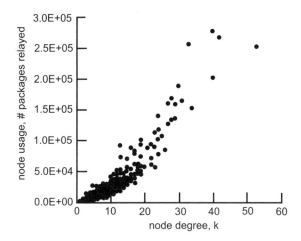

Fig 11.2   Correlation between node degree and supported traffic in the reinforced scale-free topology (1024 nodes, 2048 links), for shortest path algorithm (see text for details).

removing less than 1/5th of the population (it is assumed that nodes reached by the malicious entity cease to function as relays) has very little impact on the relative size of the largest component (average is ~0.97, SD ~0.02). Indeed, in this very robust design, the extra 1024 connections provide massive redundancy, meaning that most surviving nodes can still reach each other even when the original shortest route has been severed. So even though losing ~17% of the nodes to a virus may be considered unacceptable in itself, a valid preliminary conclusion would still be that, if the epidemic can be contained at this relatively early stage, the network as a whole would remain widely operational.

This, however, disregards the fact that geometric propagation of an epidemic cannot be assimilated to cumulative random node removal, because the very nature of scale-free topology means highly connected vertices are never far away from the initial penetration point. This in turn creates the possibility of harmful side effects, in the form of catastrophic cascade failure. In order to test this eventuality, we re-wrote routing tables after the malicious entity's progression stopped, therefore redirecting packets around missing relays. Simultaneously, we introduced the additional constraint that any node having to handle more than twice the traffic it is designed to support under normal conditions (as a result of the re-routing process) fails due to overload. As long as this event occurs at least once in a cycle (i.e. until no surviving node exceeds 200% of its original capacity), we kept re-writing the tables so as to take into account this additional damage and re-evaluated traffic distribution. Finally, we assumed that traffic generated by or targeted to nodes that have failed (including those which fell to the virus) disappears with them, which is obviously a 'best case' scenario as it reduces the load on surviving relays.

The simulation results are self-explanatory. As shown in Fig 11.3, only six out of ten numerical experiments (one per network) end up with the overall situation actually stabilising shortly after the epidemic is stopped. In the remaining four, cascade failure leads to increasingly more relays going off line, up to the point where the largest component is being reduced to less than 20% of the surviving

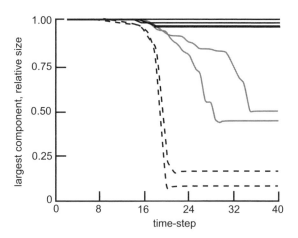

Fig 11.3　Evolution of the relative size of the largest component, before and after an epidemic (10 simulations, each one involving a different instantiation of the reinforced scale-free network) when taking into account the effects of cascade failure (see text for details). Damage stabilises after the end of the epidemic ($t = 16$) in only 6 simulations out of 10 (solid curves). In two simulations, the relative size actually falls from 90% to 10-15% due to indirect damage (dashed curves).

population in two simulations (Fig 11.3, dashed curves). In one case, the total amount of damage attributable to progressive overload (322 failures) was over 2.5 times that caused by the malicious entity itself (125). The obvious conclusion is that even rapid reaction in a topologically robust architecture (scale-free network reinforced by a secondary layer of random links) is by no means certain to successfully protect the system from total breakdown.

11.2.2   Results from the Hypergrid

The situation, however, could not be more different when the same malicious entity is attacking a network featuring hypergrid topology. In short, this architecture can be described as a tree whose 'leaves' are interconnected at random so as to feature a completely homogeneous node degree (see Fig 11.4). For information, the diameter of a hypergrid is typically slightly inferior to that of its scale-free counterpart (~9 versus ~10 in a network comprising 1024 nodes with an average degree of four), and it also increases logarithmically with size. There is of course a price to pay and the average path length (as opposed to the diameter which is the maximum) is higher in the hypergrid (~5.7 versus ~4.9 in the scale-free network, see Fig 11.5). Yet as shown in Fig 11.6, the most striking difference between those two topologies is the

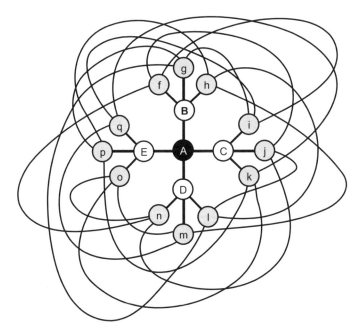

Fig 11.4   Schematic representation of a hypergrid of degree 4. All nodes have 4 first neighbours, those non-hierarchical links on the periphery are cross-allocated at random.

distribution of traffic, concentrated around the average in the hypergrid due to the absence of just a few highly connected nodes acting as primary relays.

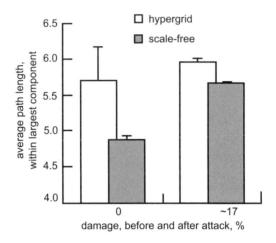

Fig 11.5 Comparison of the average path length, before and after attack, for two networks of the same size (1024 nodes, 2048 links). One is the reinforced scale-free architecture, the other is the hypergrid.

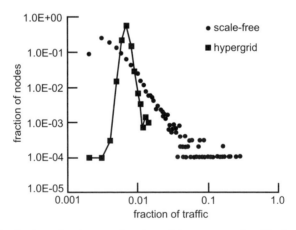

Fig 11.6 Distribution of nodes as a function of the amount of traffic they support in the intact network. For the reinforced scale-free topology, it follows the power-law typical of this type of architecture (as can be expected from the correlation shown on Fig 11.2). For the hypergrid, it is much more homogeneous.

This completely different design philosophy has two major implications for epidemic control. Firstly, there is no single point (central hub) that, once reached, provides simultaneous access to a large number of potential targets; secondly, the more homogeneous distribution of traffic across the network in normal conditions makes cascade failure far less probable when re-routing is needed.

Conducting exactly the same test as for the scale-free topology, it now requires 21 time-steps (versus 16) to reach the same amount of damage (~17% of nodes infected), this slower propagation being attributable to the more decentralised structure of the system (self-replication rate is still 0.1). Also note (see Fig 11.3) that the average path length degrades more gracefully as damage increases. But even more importantly, in none of the 10 simulations did a single relay crash as a result of traffic having to be re-routed around the compromised region (i.e. no cascade failure). This is because a more homogeneous load in normal conditions leads to less fluctuation during a crisis situation, in terms of the relative increase of traffic through surviving nodes (i.e. crossing the 200% threshold is a considerably less likely event in the damaged hypergrid).

11.2.3   Analysis

The obvious conclusion to this first set of numerical experiments is that, although reinforced by random links, the network produced by preferential attachment is only robust on a very abstract level (resilience of the largest component). Indeed, even when it does not specifically target the most highly connected nodes, the topology itself makes it likely that a propagating malicious entity will reach at least some critical hubs in just a few hops, which confirms the results of Pastor-Satorras and Vespigianni [8]. Even more worrying is the fact that such hierarchical architecture would often suffer massive cascade failure as a consequence of a relatively small fraction of relays being knocked out by the attacker.

One can only regret that an alternative design like the hypergrid, which seems to offer considerable advantages in that respect at apparently no extra cost (except for more complex routing procedures perhaps), has not been envisaged. However, until the opportunity arises to deploy a new breed of networks, which may be sooner than many expect (e.g. *ad hoc* connectivity, grid computing and their specific requirements), the question has to be: 'How do we protect the architecture we have, despite its built-in vulnerabilities?'

11.3   Adaptive Security and Epidemic Control

11.3.1   Introduction

Although the previous section hopefully contributes to putting network robustness into a new perspective, by adding some rarely discussed yet eminently practical considerations to the debate, it is mainly intended as a background to a more detailed presentation of our approach to adaptive IT security.

Relatively few research efforts in this field explicitly acknowledge the necessity to switch to some sort of decentralised management of defensive measures in order to cope with changing networking practices. The most advanced papers in that respect would be those by Ioannidis et al [12] about distributed firewall, and by Zhang and Janakiraman [13] which describes the Indra intrusion detection system. However, even those pioneering works fall short of taking into account some key aspects of variable network topology. In practice, both protocols rely on positive detection and successful localisation/identification of the security breach. In addition, the distributed firewall does not really address the problem of harmful behaviour from an authenticated access point. On the contrary, Indra is an extremely aggressive IDS that does not appear to have any tolerance for transient unauthorised activity, which is potentially very dangerous in a highly dynamic environment (false positive). Far from discarding these previous attempts though, the 'adaptive network perimeter' presented in this chapter is actually building on their respective strengths while simultaneously trying to correct their weaknesses. Indeed, in our view, the key to successful distributed defence resides in replacing any form of explicit threat notification (even decentralised) with periodic updates of the local security profile, controlled by inhibitory signalling.

Despite the fact that it is known to be a fairly inaccurate rendering of most real life systems, the square lattice is still commonly used to model distributed phenomena, including virus propagation. Most recent improvements in that respect actually find their origin in the field of theoretical physics (see, for example, Pastor-Satorras and Vespigianni [8], but also Moore and Newman [14]), and have only come to be recognised by the wider community as offering a more realistic alternative over the last few years. However, because it is a powerful demonstration tool and provides a good opportunity to explain what is meant by adaptive network perimeter (and to show how it could be implemented by using local signalling only), we will first present results obtained in this basic, grid-like architecture. Subsequently, the dynamics of contagion and epidemic control will be examined in other types of network, making it clear that the defensive strategy we propose can successfully be used in less idealised environments.

11.3.2   Adaptive Network Perimeter — Inspiration and Dynamics

The biological world offers many examples of organisms that have evolved efficient ways of dealing with a dynamic threat, either on their own or by exploiting co-operative effects that can arise in communities of interacting units (cells or individuals). For example, the immune system not only recognises pathogens, but can also focus its efforts on the place of maximum danger through a fully decentralised process (inflammatory response). Similarly, defensive recruitment in social insects can trigger an adapted reaction by attracting the exciting individuals in the vicinity of the intruder to the point where it is overwhelmed by a swarm of aggressive workers (see Millor et al [15]). The adaptive perimeter proposed uses a similar logic in order to increase the plasticity of network security systems, enabling them to react to topological changes so that defensive measures are always concentrated at the periphery. Essentially it can be viewed as being an adaptive firewall system, kept 'dormant' in nodes that are located in a safe environment, but spontaneously building up to full strength as soon as the device on which it is running is no longer suitably protected.

In a square lattice, each node is obviously connected to the same number of first neighbours ($N = 4$). The strength of the defensive measures (hereafter referred to as a 'firewall' even though it could include other types of protection) implemented on each node is measured by a real number (x) between 0 (no security checks) and 1 (maximum alert level).

In a real system, this value should be mapped on to a set of predefined security stances, ranging from very permissive to very cautious.

The adaptive response is based on the exchange of 'beacon' signals between first neighbours. These signals consist of a recognisable ID and current security level of the sender (firewall strength), accompanied by a summary of its internal state. This last component is introduced as a way of identifying those previously trusted nodes that have been compromised. We make the assumption that this would impact on their behaviour (pattern of activity), which would translate into unexpected modifications of the internal state, therefore resulting in a corrupted signature.

Upon receipt of its neighbours' beacon signals, each node computes a new alert level for itself on the basis of its current status and the information contained in the N (or less) authenticated signal packets, following a simple differential equation:

$$\frac{dx}{dt} = \frac{x(1-x)}{N}\left(N - n + \alpha \sum_{i=1}^{n} x_i\right) - \beta x \qquad \text{...... (11.1)}$$

In equation (11.1), $n \leq N$ is the number of first neighbours for which this node has received a properly formatted beacon signal, i.e. including a recognisable tag identifying the sender as a trusted member of the community. The sum then represents the security level of the n trusted neighbours $(1-x)$ standing for saturation effects. The right-hand βx term (where $0 < \beta < 1$) introduces a form of decay whereby firewall strength spontaneously lowers if not compensated. It should be noted that since there is no independent term, $x = 0$ is always a trivial solution of equation (11.1), meaning that if security is non-existent, it requires an external intervention to 'ignite' the firewall.

Examining limit cases provides useful information about system behaviour generated by equation (11.1). For example, considering the situation where none of the N first neighbours are trusted nodes (the device is isolated in the middle of a potentially hostile environment), n is equal to zero and the sum is null. Equation (11.1) then becomes:

$$\frac{dx}{dt} = x(1-x) - \beta x = x(1-\beta) - x^2 \qquad \text{...... (11.2a)}$$

and the (stable) positive solution is:

$$x = 1 - \beta \qquad \text{...... (11.2b)}$$

In other words, provided that $x > 0$ at the start (residual security) and $\beta \ll 1$ (spontaneous extinction is relatively slow), the alert level of any isolated node will build up to a value close to 1 (maximum security).

Another interesting case is found for a network comprising only 'friendly' nodes — then $n = N$ and x is identical throughout the system ($x_i = x$). In this situation equation (11.1) becomes:

$$\frac{dx}{dt} = \frac{x(1-x)\alpha Nx}{N} - \beta x = \alpha x^2(1-x) - \beta x \qquad \text{...... (11.3a)}$$

Because one of the three solutions is obviously $x = 0$ (no independent term), equation (11.3a) shares the two others with its counterpart of the second degree:

$$\frac{dx}{dt} = \alpha x^2 + \alpha x - \beta \qquad \text{...... (11.3b)}$$

So the stable and unstable solutions of equation (11.3a) are simply given by:

$$x = \frac{\alpha \pm \sqrt{\alpha^2 - 4\alpha\beta}}{2\alpha} \qquad \text{...... (11.3c)}$$

Those solutions only exist if $\alpha > 4\beta$, in which case the lower one acts as a threshold above which spontaneous decay cannot compensate for the combined self- and cross-excitation among the nodes and the entire population goes into full alert (stable solution). Since in this scenario all devices are assumed to be trustworthy, this is obviously a pathological situation that should be prevented by careful selection of the parameter values (α and β) and of the initial firewall strength (x_0). Figure 11.7 shows the variation of the excitation level (dx) as a function of firewall strength (x) for a particular configuration involving selected values of α and β.

In practice, choosing parameter values so that $\alpha < 4\beta$ (out of the range where the analytical solutions given by equation (11.3c) are real and lie between 0 and 1) is a convenient way of preventing the pathological situation presenting itself. Indeed, in this case, only the trivial solution $x = 0$ stands, and a community of mutually trusting nodes cannot 'go paranoid' — whatever the perturbation, they will always revert to a low security state. However, given the fact that the subsequent ability of nodes to quickly raise a firewall again is dependent on their latent security level, it is advisable to 'artificially' keep this one above the initial value x_0.

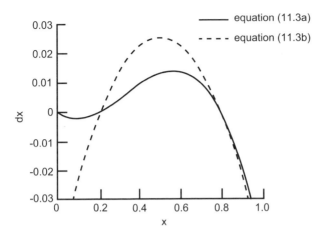

Fig 11.7   Variation of the alert level (dx) as a function of its homogenous value (x) in the 'all friends' square lattice scenario (parameter values are $\alpha = 0.3$, $\beta = 0.05$, $N = n = 4$, see text for details).

11.3.3   Static Defences versus Adaptive Network Perimeter (Square Lattice)

The following example considers the case of a 32×32 square lattice with periodic boundary conditions (torus), thus identical in size to the network used to test the influence of cascade failure (1024 nodes, twice as many links). All interactions (contamination and exchange of alert beacons in the adaptive defence scenario) take place exclusively between a node and its first four neighbours (no diagonal links). Because of the topology of this grid-like architecture, the propagation of the malicious entity is much slower than in a scale-free or random structure though, for precisely the same reason that made the hypergrid more resilient as well (homogeneous node degree). However, in this case, the effect is stronger due to the absence of short-cuts (long-range connections make networks more vulnerable — see section 11.3.4 for details). Figure 11.8 shows the progression of the epidemic in the square lattice, compared with that observed in the reinforced scale-free topology (no cascade effects), in the absence of any protection mechanism.

To test the influence of defensive measures on the propagation of a malicious entity, we have to define an 'interception' model, describing how the defender (firewall) and the attacker (virus) would interact locally (i.e. when a protected node is directly targeted). In order to keep it simple, and easy to implement in the simulation, a probabilistic approach is used — whenever a relay is under attack, a test is performed to determine whether the attempt is successful or not. If the returned pseudo-random number ($0 \le x < 1$) is lower than the node's alert level, the

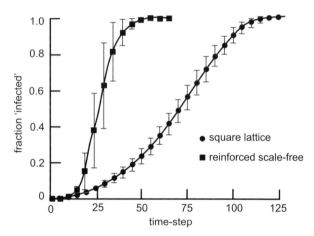

Fig 11.8   Propagation of the same malicious entity in the reinforced scale-free network and in the square lattice when no defences are in place. The low diameter and hierarchical structure are responsible for the faster invasion in the first topology (error bars represent standard deviation).

firewall is assumed to have intercepted the harmful transmission (and no contamination occurs).

It could be argued that this is not realistic, as the response of a security protocol to a given type of attack is typically 'all-or-nothing' (either the loophole has been identified and patched or it has not). We think on the contrary that a probabilistic model is a suitable way of introducing the uncertainty factor into the simulation. It is a well-known fact that in most outbreaks of computer viruses, the threat is detected and the solution made available before the epidemic reaches its peak. However, the update procedure itself is less than perfect, because it involves complex organisations with considerable momentum and is also subject to human error/ interpretation. The result is that, in effect, the probabilistic model is a realistic one. Indeed, an alert level of 0.8 does not have to mean that there is an 80% chance for a security system to successfully intercept the propagating entity. What it represents is rather that there is an 80% chance that the appropriate defensive measures are actually up and running where and when the attack effectively occurs.

We first simulated the propagation of the malicious code in a static defence scenario. There is again a single entry point, but in this case, it is assumed to be surrounded by secure gateways standing a high chance of intercepting an intrusion attempt (alert level of the infected node's first neighbours is 0.95). This situation is very similar to that of a corporate intranet, connected to the outside world via a few protected access points. Results are shown in Fig 11.9. The fact that the entry point is initially surrounded by high profile defences delays the outbreak and increases noise level (the virus keeps 'probing' the firewall, but is unsuccessful until one of its attempts happens to coincide with a lapse in security). However, the bottom line is

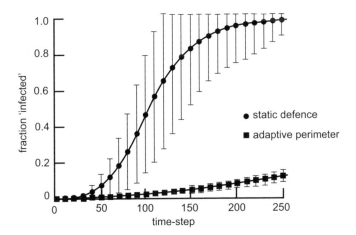

Fig 11.9 Propagation of the same malicious entity in the square lattice. In one case, a static protection system is in place (initial access point surrounded by firewalls), while in the other, the adaptive network perimeter is responsible for activating defences. The high variability in the first scenario is attributable to the fact that breaking through the firewall is a very low probability event. As a result, the epidemic's start can vary a great deal from one simulation to the other. However, as soon as the defences have been breached, the propagation is extremely fast and follows a predictable pattern. In the adaptive perimeter scenario, the progress of the epidemic is much slower because new layers of defence are constantly being activated (see text for details).

that, once inside, the epidemic progresses at the same rate as if no protection was ever in place after the 'hard' shell has been breached (whatever time this may take); the self-replicating malicious code can spread with total impunity in the 'soft' interior.

Alternatively, network defence could rely on adaptive perimeter techniques. In that case, the alert level of the four gateways surrounding the entry point would result from them receiving only three identifiable beacons out of four neighbours. Just as for the propagating malicious entity, updates are conducted 'simultaneously', and therefore, in effect, changes in the security profile between times t and $t + 1$ are computed on the basis of the information exchanged between first neighbours at time t. For parameter values $\alpha = 0.2$ and $\beta = 0.05$, the resulting alert level (minimum allowed $= 0.1$) would stabilise around $x \approx 0.86$ for the first layer of defence (i.e. lower than in the static firewall scenario, indicating weaker protection). Yet the interaction between mutually trusting relays also results in the first neighbours of those nodes in direct contact with the threat adopting an intermediate security profile (instead of the default low-alert level). Finally, as the epidemic progresses, the missing/unrecognisable beacon signals from contaminated nodes cause survivors deeper inside the once protected core to raise their own defences. The overall effect is that the propagation rate never approaches the same level as in the unprotected environment (see Fig 11.9 above), because the local security profile constantly adapts itself in order to reflect changes affecting the position and shape of the trusted domain periphery. A possible analogy would be a defensive force gradually retreating to prepared fall-back positions when put under irresistible pressure, instead of fighting a losing battle up to the point where the front line collapses (allowing the attacker to rush into unprotected territory). Figure 11.10 shows a succession of snapshots illustrating this phenomenon.

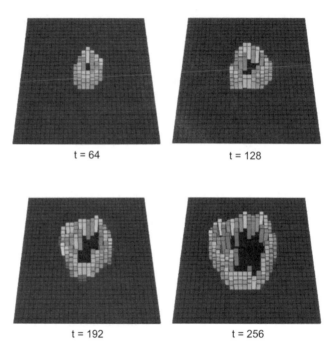

t = 64   t = 128

t = 192   t = 256

Fig 11.10   Schematic representation of the response of the adaptive network perimeter to the propagation of the malicious entity in the square lattice. The tint and height of the bars represent the alert level of the corresponding node (initial entry point is in the middle, where the infected relays are shown in black).

11.3.4   Generalisation to Other Topologies

However convincing this demonstration may seem at first, it is ultimately plagued by the fact that the square lattice is unlikely to ever become a realistic representation of any network. Even though we acknowledge and welcome novel architectures, which we think will become increasingly less hierarchical as the 'peer-to-peer' philosophy gradually replaces its 'client/server' counterpart, we also firmly believe that topology itself will never become 'flat'. Indeed, a critical aspect of network operation is that it should always be possible to reach any node from any other in only a few hops, meaning that the diameter has to grow as slowly as possible (typically as a logarithmic function of size). As a result, even advanced new designs exploiting local rules to self-organise into a coherent whole [16] are still required to feature long-range connections in order to be efficient. The drawback of course is that those same short-cuts that make the system scalable are also an ideal way for an epidemic to spread to distant, yet untouched regions.

So does adaptive network security offer a real advantage in such a 'non-Euclidean' environment, or does the existence of long-range links effectively imply that any kind of perimeter defence is meant to fail, because it will always be overrun? The answer is obvious — adaptive security can be successful if it manages to be as dynamic as the malicious entity it is trying to contain, i.e. if the alert signal can propagate faster than the virus whatever the topology. Fortunately, the adaptive network perimeter proposed does have this potential. Indeed, not only does it exploit the same channels that are available to the hostile code to replicate itself (same propagation dynamics), it also has a comfortable head start, because secondary targets (2-3 hops inside the trusted domain) typically live in 'pre-alert' state. In short, it does not really matter what the network map looks like, as long as every node knows its first neighbours, collects beacon signals from all of them, and keeps its own security profile up to date on the basis of this information.

In a hierarchical architecture (i.e. featuring heterogeneous node degree), it requires a little bit of tuning. Indeed, one compromised first neighbour is as much a threat to a hub with 100 links as it is to a local relay with only four (actually, even more so from a global point of view), and should therefore trigger the same reaction in both cases. This is easily achieved by putting an upper limit to N and only taking into account the n most-worrying beacons (plus eventual missing ones) when solving equation (11.1).

It could even be beneficial to make other parameter values (α and/or β) a function of node degree, so that highly connected relays are actually more responsive to alert signals than their less critical (lower degree) counterparts. Yet even without this extra refinement, the performance gap between static defences and adaptive perimeter remains the same, as simulation results clearly demonstrate.

This test stimulated an epidemic in the reinforced scale-free architecture. As before, we selected one node out of 1024 as the single entry point for the virus into the network. In one case, we surrounded it with static barriers (all of its first neighbours' alert levels were set to 0.95), in the other, we allowed dynamic interactions to bring the system to equilibrium before the epidemic starts. This adequately represented the situation where a firewall or the adaptive network perimeter has been set up prior to the outbreak. In the second case, the upper limit for N is set to 4, obviously identical to that dictated by the topology in the square lattice (other parameter values are $\alpha = 0.2$ and $\beta = 0.05$). Figures 11.11 and 11.12 show the evolution of the frequency distribution with respect to the surviving fraction (64 classes) as a function of time, in the static and adaptive perimeter scenario respectively.

In the first case (static defence), transition is very fast with between 100% and <2% survivors, indicating rapid propagation of the virus as soon as the firewall is breached (few intermediate values). On the contrary, the epidemic is much slower and its progress more gradual in the adaptive defence scenario because dynamic update of the alert level means that malicious code never reaches a region where it can spread with total impunity.

The obvious conclusion to this section is that in most cases, independently of the underlying topology, the adaptive network perimeter will easily outperform static defences in the long run if properly configured. In fact this is not at all surprising if one takes the secure gateway of the traditional security approach for what it really is — a single point of failure.

In this scenario, as soon as the malicious entity has detected a weakness in the firewall configuration of its target network, it has won the war. On the contrary, in the protocol we propose, it would have to fight again for every single node it is attempting to infect or hijack.

Local vulnerability may be marginally increased (lower stable alert level for those elements in direct contact with the threat), but the system as a whole resists much better because defensive capabilities are not limited to those nodes making up the original periphery of the network.

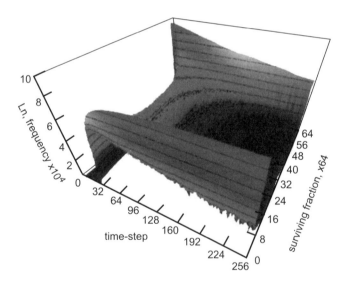

Fig 11.11   Evolution of the frequency distribution with respect to the surviving fraction as a function of time, in the reinforced scale-free network, protected by static defences (104 realisations). The U-shaped distribution reflects the rapid transition toward the fully infected state once the firewall has been breached (the pattern is similar to that observed in the square lattice, with the exception of the time-scale).

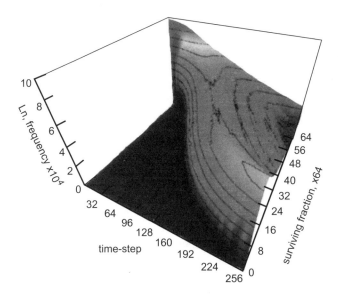

Fig 11.12 Evolution of the frequency distribution with respect to the surviving fraction as a function of time, in the reinforced scale-free network, protected by the adaptive perimeter protocol (104 realisations). The much smoother transition results from the slower progression of the epidemic, due to the spontaneous build-up of new defences each time one layer has been breached by the malicious entity. Also, note the completely flat area on the left-hand side, indicating that not even one of the 104 simulations saw any significant damage being caused to the network in the early stages of the attack.

11.4   The 'Big Picture'

11.4.1   The Future of Computing and its Security Implications

As already mentioned, we are convinced that future distributed architectures will see increasingly more non-hierarchical interactions taking place between constitutive units. From our point of view, reversing that trend is neither possible nor desirable. Not only does it open fascinating perspectives in terms of better resource management [17, 18], but also has very positive implications for network robustness and survivability, as previously discussed in the case of the hypergrid.

So far, we have shown that a 'decentralised' topology (i.e. featuring homogeneous node degree) is less prone to cascade failure and therefore better equipped to maintain quality of service when damaged than a hierarchical architecture. We have also presented strong evidence that, in theory, an adaptive perimeter defence protocol is considerably more capable than the traditional firewall strategy when it comes to resisting a dynamic threat like a propagating malicious entity. But the real challenge and the main purpose of the work described in this chapter is to demonstrate that these two ways of improving network survivability are actually complementary, and are most successful when combined into an integrated approach to information security.

As traffic and/or processing is increasingly more evenly distributed, large systems become less sensitive to a directed attack, which can be regarded as the trivial counterpart to the results of Albert et al [5] on scale-free networks. So if future designs do indeed belong to this category, this particular threat is likely to progressively fade away. On the other hand, however decentralised distributed systems effectively become in the coming years, they will always have to include the one critical feature that makes them scalable — small diameter and average path length. If the network is to be non-hierarchical, this vital property can only come from increasing the number (and range) of connections between those nodes that used to be at the lowest level in the organisational chart, together with their access privileges. Indeed, this active contribution of every member device to the smooth operation of the system as a whole is probably the most important design rule in the popular *ad hoc* network concept (see, for example, Johnson [19]).

But this obviously comes with a security warning attached. If every single node is empowered with functions that traditionally belong to specialised relays, and is therefore entitled to receive, process and transfer all kinds of information, the danger of a malicious entity finding a way of causing nuisance from a seemingly insignificant position increases dramatically. In short, the constraints of networking dictate that removing the possibility of knocking down the system with a single blow automatically provides the potential attacker with more ways to inflict less critical damage which, in the long run, can turn out to be just as harmful. Therefore, eliminating the 'single point of failure' by improving the architecture's versatility has to be accompanied by some serious thinking into how to achieve better protection of individual nodes in a dynamic environment, which of course is what the adaptive perimeter is all about.

11.4.2   The Adaptive Hypergrid Perimeter

The ultimate goal in the security of distributed systems has to be achieving overall robustness to accidental breakdown (which includes prevention of cascade failure), while at the same time being able to withstand both directed and distributed attacks, all of this without compromising efficiency under normal operating conditions.

Even though the adaptive network perimeter is considerably more successful than static defences in protecting the reinforced scale-free network (see Fig 11.12), in extreme stress conditions, it cannot help prevent indirect damage caused by re-routing through less-capable nodes (cascade failure). If being able to shield a larger fraction of the population from the epidemic is obviously a necessary condition to ensure network survivability, it is only part of any global answer. Indeed, the efficiency of the adaptive defence protocol with respect to slowing down the progress of the malicious entity is no guarantee that the system remains operational as a whole, because there is no relationship between having succeeded in repelling a virus assault and surviving a traffic surge. In other words, a node that has raised appropriate security barriers in time to avoid being infected can still succumb to the overload resulting from some of the traffic being re-routed around other relays which have been caught 'off-guard'. Ultimately, the only difference with the scenario developed in section 11.2 'epidemic propagation and cascade failure', is the time-scale. Where it takes only 16 time-steps before the virus has reached enough nodes for the cascade failure to 'finish the job' with no defences in place, our simulations show that it does require around eight times more (~128) in a system protected by the adaptive protocol.

But why not combine the hypergrid's ability to absorb widespread re-routing of diverted traffic with the adaptive network perimeter's efficiency in slowing down the progress of an epidemic? This way, the system can benefit on two fronts simultaneously — exposed nodes become more likely to adopt a security profile allowing them to successfully ward off an attack, while topological changes resulting from the occasional success of the malicious entity (compromised relays) are less likely to threaten system-wide operations.

Again this was tested by simulating a network comprising 1024 nodes and twice as many links, submitted to attack in the form of an on-going epidemic. The numerical experiment's protocol is therefore similar to that described in section 11.2, with the exception that the adaptive defence protocol presented in section 11.3 'adaptive security and epidemic control' is now active. Also, instead of allowing the virus to propagate for a given amount of time, we stop this phase of the numerical experiment as soon as 25% of the nodes have been successfully infected, which previous tests suggest has ~50% chance of triggering cascade failure in the reinforced scale-free network. As shown in Fig 11.13, in both types of network, at the end of the epidemic phase, most of the nodes are on either 'medium' (~0.7) or 'high' (~0.9) alert. This indicates that they have successfully 'detected' the presence of a threat, and adjusted their security profile accordingly (i.e. reached the stable state that is most appropriate to their own individual location with respect to the origin of the pertubation).

However, this does not prevent the system from subsequently collapsing under the influence of local traffic surges in the predicted half of the ten experiments involving the reinforced scale-free topology. In the hypergrid on the contrary, no cascade failure occurred, resulting in the total surviving population stabilising right at the end of the epidemic (no overload-related crashes, see Fig 11.14). Moreover, virtually all uninfected relays are still included in the largest component at the end of the simulation — only a minute fraction (<1%) of the successfully protected nodes are effectively cut off from the network (and from any shared resource), against a massive average 40% in the other scenario. So the adaptive hypergrid perimeter may well point the way to a holistic approach to network design and defence — one that will eventually lead to the development of dynamic, scalable and decentralised, yet also resilient and secure, distributed systems.

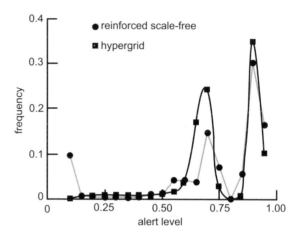

Fig 11.13   Distribution of surviving nodes as a function of their alert level in the reinforced scale-free network and in the hypergrid, both protected by the adaptive perimeter protocol, after 25% of the total population has been infected.

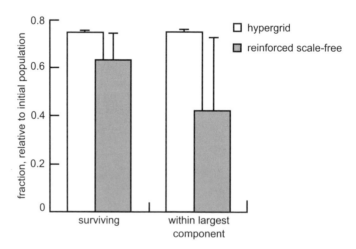

Fig 11.14   Comparison of the fraction of nodes having survived both the epidemic and the resulting cascade failure, in the reinforced scale-free network and in the hypergrid, both protected by the adaptive perimeter protocol (propagation of the virus is still stopped when 25% of the population has been infected). It should be noted that only in the case of the hypergrid does the largest component include virtually all of the survivors (see text for details).

11.5   Summary

This chapter has discussed network security and robustness as a single, intricate problem. Although it may be argued that boosting the defensive capabilities of distributed systems and increasing their operational stability are different agendas in terms of the techniques involved, it is our opinion that they should not be separated. Indeed, ultimately, they both aim at the same target — improving network survivability so as to ensure that acceptable quality of service can be maintained under stress, whether accidental or deliberate. As systems regrouping large numbers of interconnected devices become simultaneously more complex and dynamic, attack strategies based on exploiting architectural vulnerabilities in order to cause damage far beyond originally targeted nodes will become increasingly common. In a sense, the infamous 'distributed denial of service' attack, whereby a pool of hijacked devices are enrolled in an attempt to flood a network with illegitimate service requests, can be regarded as an early example of this tactic being put into practice. We argue that the inextricable relationship between security and architecture has to be much more widely accepted on the defenders' side if we are to devise suitable ways of protecting what is rapidly becoming a planetary, 'always-on' information processing entity whose development (or blueprint) cannot be controlled. We think we have given enough evidence that adaptive perimeter techniques can successfully be combined with inventive design in order to reach a level of resilience to attack and possible side-effects far superior to that achievable by following either of these routes.

In addition, we would now like to present arguments in favour of the dynamic defence paradigm, this time as an alternative to the ubiquitous security (also known as 'paranoid') approach. Indeed, running a personal firewall on every single piece of equipment, while simultaneously promoting the use of VPNs on an even larger scale, may seem like the most direct solution to the polymorph threat faced by dynamic architectures. Our opinion, however, is that, in the long run, this option will turn out to be utterly impractical, because in essence it is trying to cope with a change in networking philosophy by finding *ad hoc* 'tricks' to keep conventional security going.

What the overwhelming majority of users want (and the driving force to most of the research going on in the distributed computing area) is seamless connectivity. The objective is quite simply to be capable of exchanging information and requesting services from a variety of 'plug-and-play' devices that do not need to go through a painful synchronisation and authentication process each time they join one of many networks. Most people will consider taking security precautions only as far as they have a negligible impact on the ease of use of such systems. As a result, even today, only a minute fraction of those users who could benefit from remotely accessing distributed services actually do. They quite legitimately refuse to accept the extra hassle of having to comply with those countless requirements that porting conventional security procedures into the world of dynamic networking imposes. Now in the near future, as the need for immediately available information/ resources keeps growing, any organisation that offers this kind of alternative to its people will arguably grind to a halt. In short, the paranoid approach does not work, not because it cannot prevent security breaches (it can), but because, in practice, ubiquitous security cannot be enforced without compromising operability.

The adaptive network perimeter, on the other hand, seems to offer a solution that is much more in phase with recent advances in design principles. It is flexible and autonomous, uses logical interaction between first neighbours to regulate their alert level, and distributes security responsibilities across the system. But even more importantly, instead of relying on one-off authentication and explicit threat notification, it exploits continuous cross-inhibition within a population of mutually trusting nodes to adjust security profiles on the fly. The result is a completely different way of dealing with intrusion and/or changes in domain topology. When a device initiates a connection to the network, it is not denied any service by default and then, after having been successfully authenticated, never checked again; rather, its neighbours continuously adapt their profile, either progressively adopting a more 'relaxed' attitude (if the newcomer is recognised as trustworthy) or, on the contrary, switching on increasingly more defence systems (if it is not).

This potentially means a substantial reduction in the security-related overhead, as a device only involved in benign interactions with trusted neighbours can obviously afford to lower most of its barriers, which in turn translates into less hassle for the legitimate user. And there is an extra benefit for the paranoid network administrator as well — because updates are continuous and rely on inhibitory signalling, a trusted device that starts exhibiting abnormal behaviour will also trigger an 'inflammatory' response which is the key to repelling attacks by insiders.

In the end, what we propose is a security protocol that is simultaneously less aggressive and more 'intelligent'. It does not ring the alarm bell and freeze all processes as soon as it comes across something unusual, which in the huge majority of cases will turn out to be the right decision. Indeed, in an increasingly dynamic environment, many connections initiated by unidentifiable devices will have no hostile motives (they may belong to visitors or to legitimate users who simply omitted to register them). On the other hand, the adaptive perimeter will avoid signing a blank cheque to anybody carrying an authorised device with the right access codes, reacting immediately if, for example, a potentially harmful application (port-scanner, suspect screen saver, etc) is turned on. As a whole, this attitude, which comes down to giving any node that wants to join the benefit of the doubt, is probably the key to network security in the 21st century. Indeed, granting the newcomer immediate access, while at the same time ensuring that defensive measures in its neighbourhood spontaneously but temporarily build up (so as to be able to face any eventuality) is the best way to reconcile seamless connectivity with acceptable security and privacy.

The main objective of this work is to promote a vision of information insurance that fits in to the bigger picture and does not disregard structural changes to focus exclusively on minor case-by-case adjustments. When the Internet took off in the mid-1990s, an evolutionary process started that is literally turning networks inside out — no more fixed boundaries to protect, no more convenient access points from where to conduct all security checks. What many surveys actually show is that our planetary network has become less like a man-made structure with pre-planned design features, and more like an ecosystem obeying its own, complex dynamics. Saying that the Internet is 'alive' is nonsense, despite what has been suggested in popular science magazines (see Brooks [20]). Saying that it is 'life-like', on the other hand, is common sense. As a result, most things in distributed computing have to go through a mutation process just to stay in the game, and network security is no exception. Only in this case, the stakes are particularly high, as failure may suddenly turn the dreams of a golden age for information technology into a cyber-punk nightmare of post-apocalyptic chaos.

References

1   Forrest, S., Hofmeyr, S. and Somayaji, A.: *'Computer Immunology'*, Comms of the ACM, **40**(10), pp 88-96 (1997).

2   Kephart, J. O. and White, S. R.: *'Directed-graph Epidemiological Models of Computer Viruses'*, Proceedings of the IEEE Computer Society Symposium on Research in Security and Privacy, pp 343-359 (1991).

3   Kephard, J. O.: *'How Topology Affects Population Dynamics'*, in Langton, C. (Ed): *'Artificial Life III: Studies in the Sciences of Complexity'*, pp 447-463 (1994).

4   Erlanger, L.: *'21st Century Security'*, Internet World Magazine — http://www.internetworld.com/magazine

5   Albert, R., Jeong, H. and Barabási A. L.: *'Error and attack tolerance of complex networks'*, Nature, **406**, pp 376-382 (2000).

6   Callaway, D. S., Newman, M. E. J., Strogatz, S. H. and Watts, D. J.: *'Network Robustness and Fragility: Percolation on Random Graphs'*, Phys Rev Lett, **85**, pp 5468-5471 (2000).

7   Cohen, R., Erez, K., ben-Avraham, D. and Havlin, S.: *'Resilience of the Internet to random breakdowns'*, Phys Rev Lett, **85**, pp 4626-4628 (2000).

8   Pastor-Satorras, R. and Vespigianni, A.: *'Epidemic spreading in scale-free networks'*, Phys Rev Lett, **86**, pp 3200-3203 (2001).

9   Strogatz, S. H.: *'Exploring complex networks'*, Nature, **410**, pp 268-276 (2001).

10   Faloutsos, M., Faloutsos, P. and Faloutsos, C.: *'On Power-Law Relationships of the Internet Topology'*, ACM SIGCOMM '99, Comput Commun Rev, **29**, pp 251-263 (1999).

11   Barabási, A. L. and Albert, R.: '*Mean-field theory for scale-free random networks*', Physica A, **272**, pp 173-187 (1999).

12   Ioannidis, S., Keromytis, A., Bellovin, S. and Smith, J.: '*Implementing a Distributed Firewall*', Proc of the 7th ACM Conference on Comp and Comm Security (CCS) (2000).

13   Zhang, Q. and Janakiraman, R.: '*Indra: A Distributed Approach to Network Intrusion Detection and Prevention*', Technical Report WUCS-01-30, Washington University, St Louis (2001).

14   Moore, C. and Newman, M. E. J.: '*Epidemics and percolation in small-world networks*', Phys Rev, **E61**, pp 5678-5682 (2000).

15   Millor, J. M., Pham-Delegue, M., Deneubourg, J. L. and Camazine, S.: '*Self-organised defensive behavior in honeybees*', Proc Natl Acad Sc, **96**, pp 12611-12615 (1999).

16   Bonsma, E.: '*Fully decentralised, scalable look-up in a network of peers using Small World Networks*', Proc of the 6th World Multi Conference on Systemics, Cybernetics and Informatics (SCI2002), pp 147-152 (2002).

17   Foster, I. and Kesselman, C.: '*Computational Grids*' in Foster, I. and Kesselman, C. (Eds): '*The Grid: Blueprint for a New Computing Infrastructure*', Morgan-Kaufman (1999).

18   Foster, I., Kesselman, C. and Tuecke, S.: '*The Anatomy of the Grid: Enabling Scalable Virtual Organizations*', Intl J of High Performance Comp App, **15**, pp 200-222 (2001).

19   Johnson, D. B.: '*Routing in ad hoc Networks of Mobile Hosts*', Proc of the Workshop on Mobile Computing Systems and Applications, pp 158-163 (1994).

20   Brooks, M.: '*The Global Brain*', New Scientist, **2244**, pp 22-27 (2000).

ACRONYMS

3GPP	Third Generation Partnership Project
ABM	agent-based modelling
ADM	add drop multiplexer
ADSL	asymmetric digital subscriber loop
AIS	artificial immune system
ANIP	access network improvement programme
AS	autonomous system
ASDH	access synchronous digital hierarchy
ATM	asynchronous transfer mode
BCM	business continuity management
BN	Bayesian network
BS	base-station
CAPM	capital asset pricing model
CB	competent body
CC	command and control
CDF	cumulative distribution function
CENELEC	Comité Européen de Normalisation Électrotechnique (European Committee for Electrotechnical Standardisation)
CFA	confirmatory factor analysis
CISPR	International Special Committee on Radio Interference
CLJ	commitment and loyalty to job
COTS	commercial off the shelf
CRM	customer relationship management
DAG	directed acyclic graph
DP	distribution point

DWDM	dense wavelength division multiplexing
EC	European Commission
EGT	evolutionary game theory
EMC	European electromagnetic compatibility
EMCAS	electricity market complex adaptive system
EMU	European monetary union
EN	Euro Norme
EPoP	European point of presence
EV	enterprise venturing
FDDI	fibre distributed data interface
FDTD	finite difference time domain
FRU	field replaceable unit
FSM	finite state model
GUI	graphical user interface
HCM	human centred management
HMM	hidden Markov models
ICRM	intelligent customer relationship management
ICT	information and communication technology
IEC	International Electrotechnical Commission
IP	intellectual property/Internet protocol
IRR	internal rate of return
ISP	Internet service provider
ITE	information technology equipment
Know SF sd	knowledge scale factor standard deviation
Know SF	knowledge scale factor
Know th sd	knowledge threshold standard deviation
Know th	knowledge threshold
LAN	local area network
LRD	long-range dependent
MoM	method of moments
NE	network element
NMS	network management system

NPV	net present value
OCM	open communication from management
JEC	(Official) Journal of the European Commission
OSPF	open shortest path first
PACS	planning, assignment and configuration system
PCP	principal connection point
PCT	patent co-operation treaty
PDA	personal digital assistant
PDF	probability density function
PDH	plesiochronous digital hierarchy
PIR	passive infra-red
PoP	point of presence
PSTN	public switched telephone network
PV	present value
QoS	quality of service
RA	Radiocommunications Agency
RAn	robustness analyser
RIP	routing information protocol
ROI	return on investment
RRM	radio resource management
RV	random variable
SARC	simulated annealing for restoration capacity
SD	standard deviation
SD	Systems Dynamics
SDH	synchronous digital hierarchy
SE	serving exchange
SEM	structural equation modelling
SIR	signal to interference ratio
SLA	service level agreement
SPRing	shared protection ring
STM	synchronous transfer mode
TB	transport block

TCF	technical construction file
TCO	total cost of ownership
TDM	time division multiplex
TFS	transport format set
TPEN	Transborder Pan-European Network
TSI	time-slot interchange
TTI	transmit time interval
UE	user equipment
UML	Unified Modelling Language
UMTS	Universal Mobile Telecommunications System
UTD	unified theory of diffraction
VC	virtual channel/circuit
VPN	virtual private network
WDM	wavelength division multiplexing

INDEX